ALLE ZEIT WACH 1842

L.R. Caplan H.C. Hopf (Eds.)

Brain-Stem Localization and Function

With the Collaboration of R. Besser and G. Krämer

With 116 Figures and 20 Tables

Springer-Verlag
Berlin Heidelberg New York London Paris
Tokyo Hong Kong Barcelona Budapest

Prof. Dr. Louis R. Caplan
Department of Neurology
New England Medical Center, Tufts University
750 Washington Street, Boston, MA 02111, USA

Prof. Dr. Hanns Christian Hopf
Neurologische Klinik, Universität Mainz
Langenbeckstraße 1, 55131 Mainz, Germany

ISBN-13:978-3-642-78174-2 e-ISBN-13:978-3-642-78172-8
DOI: 10.1007/978-3-642-78172-8

Library of Congress Cataloging-in-Publication Data
Brain-stem localization and function/L.R. Caplan, H.C. Hopf, (eds.); collaboration of R. Besser and G. Krämer. p.cm. Includes index.
ISBN-13:978-3-642-78174-2
1. Brain stem-Diseases. 2. Brain-Localization of functions.
3. Cerebrovascular disease. I. Caplan, Louis R. II. Hopf, H. Ch. (Hanns Christian) III. Besser, R. (Roland) IV. Krämer, Günter, Dr. med. [DNLM: 1. Brain Stem-physiology. 2. Central Nervous System Diseases-diagnosis. WL 310 B8137 1993]
RC394.B7B73 1993 616.8-dc20 DNLM/DLC 93-5073

Softcover reprint of the hardcover 1st edition 1993

Typesetting: Best-set Typesetter Ltd., Hong Kong
25/3130-5 4 3 2 1 0 – Printed on acid-free paper

Preface

The brain stem and cerebellum are structures that have fascinated and puzzled physicians, anatomists, and physiologists for centuries. Relatively small in volume compared to the much larger and more majestic human cerebrum, and hidden away in relative obscurity in the back of the head under the protection of the bony posterior fossa and the firm dural envelope, the tentorium cerebelli, these critical neural structures defied precise study during life until very recently. Recent advances in brain and vascular imaging and improved understanding of brain-stem reflex and tract functions and their measurement by electrophysiological techniques have presented an opportunity for clinicians and researchers to better study, during life, patients with stroke and other conditions that involve the brain stem and cerebellum. A congress was held at the Neurology Clinic of the University in Mainz, April 4–5, 1992, during which clinicians and researchers reported and discussed modern topical diagnosis of diseases of the brain stem and cerebellum. This volume represents the edited proceedings of that congress.

Early information about brain stem anatomy and functions came from anatomical studies by Ludwig Türk, Paul Flechsig, Rudolph von Kölliker, Karl Burdach, Vladimir Bekterew, and Benedikt Stilling, among others, all working in Germany during the middle years of the nineteenth century. Johann Reil, a German anatomist, in the beginning of the nineteenth century had described the structure and some functions of the cerebellum which he called the little brain ("*Kleinhirn*"). Clinical information about patients with brain-stem lesions did not develop until the end of the nineteenth century when clinicians and pathologists such as Hermann Weber, Moritz Benedikt, and Adolph Wallenberg reported individual patients studied clinically and at necropsy who had focal brain-stem lesions. Even today, we continue to remember these individuals by using eponymic designations such as Weber's and Wallenberg's syndromes for some focal brain-stem lesions. Much of this early work was performed in Germany. Henry Duret, Otto Heubner, Charles Foix, and James Stopford, during the late nineteenth and early twentieth centuries anaylzed and described the vascular supply of the brain stem and cerebellum. A quantum leap forward came when Jerzy Olszewski and Donald Baxter published their detailed atlas of brain-stem anatomy (*Cytoachitecture of the Human Brain Stem*) in 1954.

Walle Nauta, the neuroanatomist, the individual who was probably most responsible for first stimulating one of the editors' (LRC) interest in neurology and the nervous system, compared the brain-stem to a small, long but compact village that was situated adjacent to a large metropolis. Through this village ran large highways carrying heavy traffic to and from the city. Near these long streets were nests of villager's homes and shops. Paths, alleys, roads, and small streets connected the shops and homes and served local traffic while the super highways served through fast traffic just passing by the city. Of course there were some streets that allowed villagers also to enter these long highways. Similarly the brain-stem contains many nuclear groups, most of which are related to local functions housed within the head. These include motor nuclei which control bulbar, facial, and oculomotor muscles and sensory nuclei and pathways which receive afferent input from the face and head and the special sense organs housed therein. Autonomic visceromotor nuclei are also prevalent. There are many local tracts and pathways (e.g., the spinal tract of V, medial longitudinal fasciculus, central tegmental tract) that connect these various brain-stem nuclei. Some of the pathways subserve cranial brain-stem reflexes such as the blink, corneal, masseter, pharyngeal, and vestibulo-ocular reflexes. In addition, there are larger tracts connecting the brain-stem and its cerebellar appendage (superior, middle, and inferior cerebellar peduncles) and, most important, key afferent and efferent long, heavily myelinated tracts that connect the cerebrum, spinal cord, cerebellum and brain-stem (e.g., corticospinal and corticobulbar tracts, medical lemnisci, spinocerebellar, spinothalamic tracts etc.). All of these different type structures are contained in a very small, compact, absolutely critical for life structure which is only about the size of a cross-section of a human finger.

Brain-stem disorders until very recently defied exact analysis during life. This key structure was very small, hidden within bone and dura, and impossible to image well. Only at necropsy was it possible to define the structures involved and their pathology. For the greater part of the twentieth century the only available investigations which could clarify brain-stem and cerebellar lesions clinically were pneumencephalography and cerebral angiography. Pneumencephalograms only showed the fourth ventricle and posterior fossa cisterns and did not show changes within brain parenchyma. Angiograms only opacified the feeding arteries and draining veins. Both investigations were uncomfortable and had a small but definite risk for the patient studied. Even computed axial tomography, introduced in the 1970s, was not very helpful in defining brain-stem lesions other than hemorrhages because of bone-related artifacts and poor definition of parenchymal structures. Only very recently, with the advent of magnetic resonance techniques and improved ultrasound has it become possible to safely and effectively image the posterior fossa brain and vascular structures in vivo.

Many patients with brain-stem dysfunction, especially those with vascular occlusive disease – the most common cause of brain-stem abnormalities – have altered function before morphological lesions appear. Even in those with morphological lesions such as infarcts, there may be additional nonfunctioning tissue adjacent or distant from the infarcts. The newer imaging technologies detect altered anatomy and morphology but not abnormal physiology. Clinical neurological examination accurately defines some localizing abnormalities but other signs, for example, abnormalities of corticospinal tracts, long sensory tracts, and cerebellar signs, are difficult to localize rostrocaudally within the brain stem. Improvements in technology for measurement of evoked responses after sensory stimuli and electrophysiological study of a number of brain-stem reflex functions which have well-defined anatomies has helped further detect the localization of the brain-stem dysfunction. In Nauta's analogy to the village, the local paths and roads more accurately tell you where you are in the village than knowing that you are speeding along a highway somewhere. The combination of clinical signs, electrophysiological analysis, brain imaging, and vascular imaging now allow better definition and understanding of brain-stem and cerebellar lesions than was ever possible in the past.

The Mainz congress was attended by acknowledged world leaders in the physiology and electrophysiology of brain-stem reflex functions and clinical features of brain-stem disorder, including Professors Thomas Brandt, (Munich, Germany), Reinhard Dengler (Hanover, Germany), Johannes Dichgans (Tübingen, Germany), Volker Henn (Zürich, Switzerland), Ludwig Gurtmann (Morgantown, W. VA., USA), Dieter Kömpf (Lübeck, Germany), Francois Maugière (Lyon, France), B. W. Ongerboer de Visser (Amsterdam, The Netherlands), Franco Regli (Lausanne, Switzerland), and many others. The congress provided an opportunity for researchers and clinicians around the world to share progress in the field. Presentations focussed on: the electrophysiology, measurement, and localizing value of the various brain-stem reflex functions; common brain-stem and cerebellar disorder such as stroke; clinical brain-stem and cerebellar phenomenology; and the brain and vascular imaging of patients with disease within the posterior fossa. Ample opportunity was available for discussion and questions. The congress proved even more successful than predicted since the large conference hall could not hold all the more than 800 participants who gathered together. Satellite rooms which had video monitors of the meeting had to be used to accommodate all those present.

The editors have tried to capture the excitement of the proceeding in this small present volume. This book should stand as a fitting tribute to the sophisticated approaches of classic neurology to clinical analysis of functional impairment. We hope that the ideas, methods, and findings convince and stimulate others to emphasize dysfunction as an important guide for localization. Morphologic criteria alone, even those shown by magnetic resonance imaging, can be misleading.

Boston, MA, USA — Louis R. Caplan
Mainz, Germany — H. Christian Hopf

Contents

List of Contributors

Altenmüller, E.
Neurologische Klinik, Kliniken Schnarrenberg
Eberhard Karls-Universität
Hoppe-Seyler-Straße 3, 72076 Tübingen, Germany

Barth, A.
Service de Neurologie
Centre Hospitalier Universitaire Vaudois
1011 Lausanne, Switzerland

Besser, R.
Neurologische Klinik, Universität Mainz
Langenbeckstraße 1, 55131 Mainz, Germany

Boegner, F.
Abteilung für Neurologie, Klinikum Steglitz
Freie Universität Berlin
Hindenburgdamm 30, 12203 Berlin, Germany

Bogousslavsky, J.
Service de Neurologie
Centre Hospitalier Universitaire Vaudois
1011 Lausanne, Switzerland

Brandt, T.
Neurologische Klinik, Klinikum Großhadern
Ludwig Maximilians-Universität München
Marchioninistraße 15, 81377 München, Germany

Büttner, U.
Neurologische Klinik, Klinikum Großhadern
Ludwig Maximilians-Universität München
Marchioninistraße 15, 81377 München, Germany

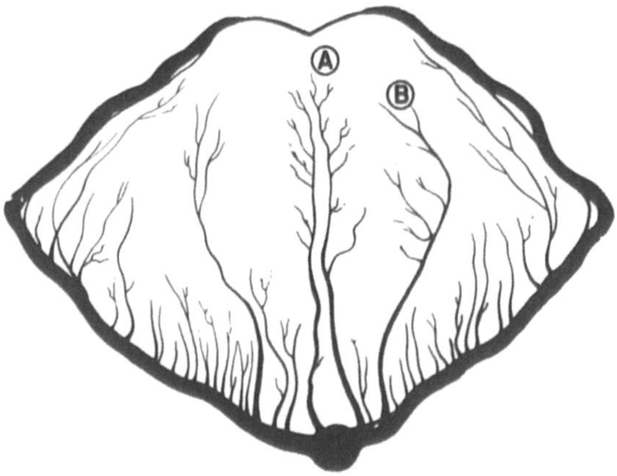

Fig. 7. Penetrating arteries of the caudal pons. *A* and *B* are median and paramedian arteries. [From 17]

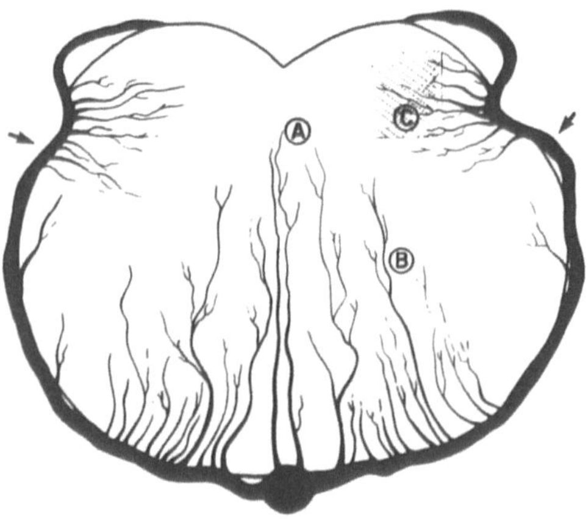

Fig. 8. Rostral pons. Circumferential arteries (*black arrows*) are shown with penetrating artery branches coursing through the lateral tegmentum. Hatched box (*C*) represents area of infarction caused by occlusion of these penetrating branches. [From 17]

farcts cause abulia and occasionally ptosis and minor motor asymmetries. The thalamic-subthalamic arteries arise from the BA apex, often with a single artery supplying branches to the bilateral paramedian mesencephalic-diencephalic junction. Unilateral infarcts cause dysmemory and bilateral vertical gaze abnormalities. Figure 10 shows such an infarct [21]. Bilateral lesions cause amnesia, vertical gaze palsies, and hypersomnolence [19, 20].

Lateral medullary and *medial medullary* infarcts can also result from atheromatous branch disease.

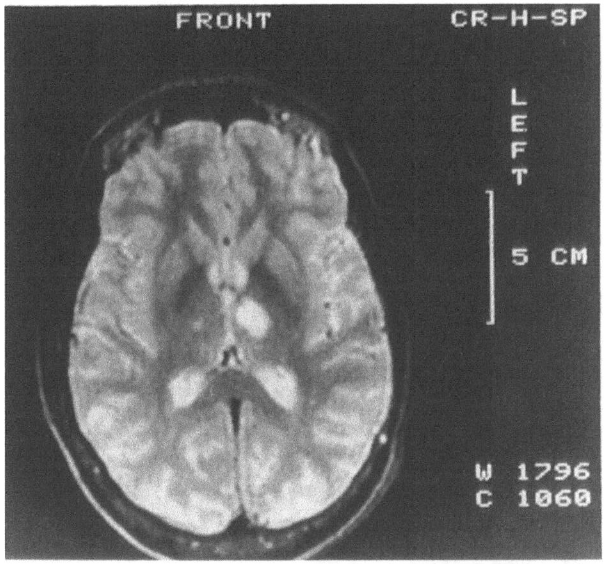

Fig. 9. MRI, T2 weighted, showing left anterolateral (polar artery territory) infarct

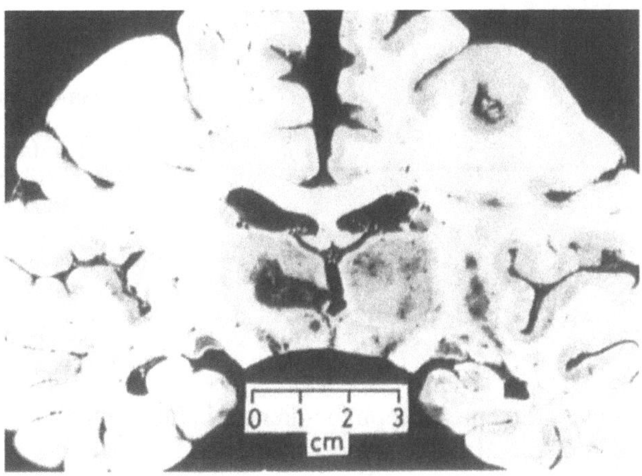

Fig. 10. Left paramedian posterior thalamic infarct in the territory of a thalamic-subthalamic artery. [From 21]

Distal BA (Top of the BA)

Embolism from the heart and the proximal posterior circulation arteries is the commonest cause of occlusion of the very distal BA [1, 2, 22]. In blacks and individuals of Chinese, Japanese, and Thai ancestry, distal BA atherostenosis occurs more often than in white men [3, 4]. Resulting infarcts

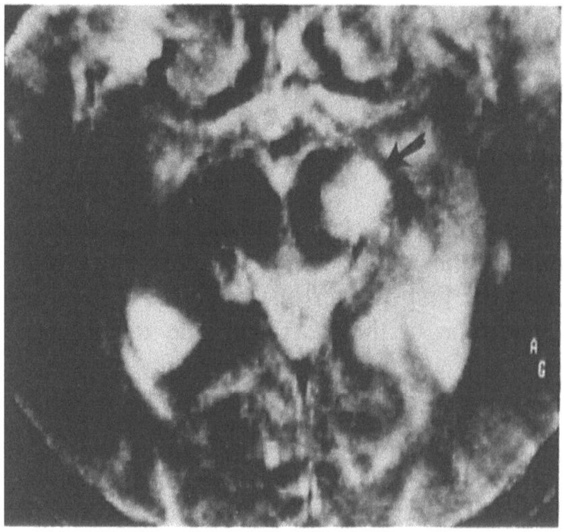

Fig. 13. MRI, T2 weighted. Infarct in the cerebral peduncle (*black arrow*) due to a proximal PCA occlusion. [From 27]

lateral thalamic and temporo-occipital lobe infarction (Fig. 12). At times the cerebral peduncle in the midbrain, supplied by the very proximal PCA, is involved [28] (Fig. 13). Occipital lobe infarcts cause a contralateral hemianopia, quadrantanopia, or homonymous scotomatous visual field defect. Hemisensory loss occurs if the lateral thalamus or the thalamoparietal sensory fibers are included in the infarct. When the most proximal PCA is occluded and infarction includes the cerebral peduncle, lateral thalamus, and posterior cerebral hemispheres, contralateral hemiplegia, hemisensory loss, and hemianopia result and the syndrome is difficult to separate from middle cerebral and anterior choroidal territory infarction.

I have reviewed briefly herein general and specific rules for correlating topographic posterior circulation brain and vascular lesions. The advent of magnetic resonance imaging and magnetic resonance angiography, extracranial and intracranial ultrasound, and neurophysiologic techniques will allow testing of these preliminary rules, which now should be considered only hypotheses or, better yet, Cider House or *Ratskeller* rules.

References

1. Caplan LR (1991) The 1991E. Graeme Robertson lecture: vertebrobasilar embolism. Clin Exp Neurol 28:1–23
2. Caplan LR, Tettenborn B (1992) Vertebrobasilar occlusive disease: review of selected aspects. 2. Posterior circulation embolism. Cerebrovasc Dis 2:256–265
3. Caplan LR, Gorelick PB, Hier DB (1986) Race, sex, and occlusive vascular disease: a review. Stroke 17:648–655

4. Gorelick PB, Caplan LR, Hier DB et al. (1985) Racial differences in the distribution of posterior circulation occlusive disease. Stroke 16:785–790
5. Caplan LR, Tettenborn B (1992) Vertebrobasilar occlusive disease: review of selected aspects: spontaneous dissection of extracranial and intracranial posterior circulation arteries. Cerebrovasc Dis 2:320–326
6. Wilkinson I, Russel R (1972) Arteries of the head and neck in giant cell arteritis. Arch Neurol 27:378–391
7. Duvernoy HM (1978) Human brainstem vessels. Springer, Berlin Heidelberg New York
8. Amarenco P, Hauw JJ, Henin et al. (1989) Les infarctus du territorie de l'artère cerebelleuse postéro-inférieure. Etude clinico-pathologique de 28 cas. Rev Neurol (Paris) 145:277–286
9. Hauw JJ, Der Agopian P, Trelles L et al. (1976) Les infarctus bulbaires. J Neurol Sci 28:83–102
10. Duffy P, Jacobs G (1958) Clinical and pathological findings in vertebral artery thrombosis. Neurology (Minneap) 8:862–869
11. Caplan LR (1983) Bilateral distal vertebral artery occlusion. Neurology (NY) 33: 552–558
12. Kubik C, Adams RD (1946) Occlusion of the basilar artery: a clinical and pathologic study. Brain 69:73–121
13. Archer C, Horenstein S (1977) Basilar artery occlusion: clinical and radiological correlation. Stroke 8:383–390
14. Fisher CM, Caplan LR (1971) Basilar artery branch occlusion: a cause of pontine infarction. Neurology (Minneap) 21:900–905
15. Caplan LR (1989) Intracranial branch atheromatous disease. Neurology 39:1246–1250
16. Helgason CM, Wilbur AC (1991) Basilar branch pontine infarcts with prominent sensory signs. Stroke 22:1129–1136
17. Caplan LR, Goodwin J (1982) Lateral brainstem tegmental hemorrhage. Neurology (NY) 32:252–260
18. Caplan LR, DeWitt LD, Pessin MS et al. (1988) Lateral thalamic infarcts. Arch Neurol 45:959–964
19. Caplan LR (1988) Vertebrobasilar system syndromes. In: Toole J (ed) Vascular diseases, part 1. Elsevier, Amsterdam, pp 371–408 (Handbook of clinical neurology, vol 9/53)
20. Bogousslavsky J, Regli F, Uske A (1988) Thalamic infarcts: clinical syndromes, etiology, and prognosis. Neurology 38:837–848
21. Bogousslavsky J, Miklossy J, Deruaz JP et al. (1986) Unilateral left paramedian infarction of thalamus and midbrain: a clinico-pathological study. J Neurol Neurosurg Psychiatry 49:686–694
22. Caplan LR (1980) Top of the basilar syndrome: selected clinical aspects. Neurology (NY) 30:72–79
23. Mehler M (1989) The rostral basilar artery syndrome: diagnosis, etiology, prognosis. Neurology 39:9–16.
24. Amarenco P, Hauw JJ (1990) Cerebellar infarction in the territory of the superior cerebellar artery: a clinicopathologic study of 33 cases. Neurology 40:1383–1390
25. Amarenco P, Hauw JJ (1989) Anatomie des artères cérébelleuses. Rev Neurol (Paris) 145:267–276
26. Pessin MS, Lathi ES, Cohen MB et al. (1987) Clinical features and mechanism of occipital infarction. Ann Neurol 21:290–299
27. Foix C, Masson A (1923) Le syndrome de l'artère cérébrale postérieure. Presse Med 31:361–365
28. Hommel M, Besson G, Pollak P et al. (1990) Hemiplegia in posterior cerebral artery occlusion. Neurology 40:1496–1499

Contribution of Magnetic Resonance Imaging to the Diagnosis of Brain-Stem and Cerebellar Infarcts

F. Regli, A. Barth, and J. Bogousslavsky

Service de Neurologie, Centre Hospitalier Universitaire Vaudois, 1011 Lausanne, Switzerland

Until recent years, most studies on the topography of brain-stem and cerebellar infarcts were based on clinicopathological correlations. Magnetic resonance imaging (MRI) has greatly contributed to the diagnosis and accurate assessment of small infarcts located in the posterior fossa [1, 3, 6–8]. In this study, we examined the topographic distribution of isolated brain-stem and cerebellar infarcts in relation to the involved arterial territories, as well as to the clinical presentation and pathogenesis of infarcts.

Patients and Methods

We studied 37 patients, 22 with an isolated symptomatic infarct in the brain stem and 15 with an isolated symptomatic infarct in the cerebellum. All infarcts were documented on MRI between 3 and 7 days after their occurrence. There were 16 women (mean age 63.8 years; range 20–83) and 21 men (mean age 63.9 years; range 24–78). Patients with multiple or pseudotumoral infarcts were excluded.

Investigations of potential causes of stroke included ECG and echocardiography in the search for a cardiac source of embolism. In patients without cardioembolism we found an arterial etiology (focal atherosclerosis of the vertebrobasilar trunk or the segmental terminal arteries, artery-to-artery embolism, microangiopathy). An accurate cause was not always determined because we did not systematically perform cerebral angiography, but all patients had Doppler sonography and MR angiographic sequences of the vertebrobasilar system.

The vascular territories in the posterior fossa were determined according to recent anatomical studies [2, 4, 5, 9].

L.R. Caplan and H.C. Hopf (Eds.)
Brain-Stem Localization and Function

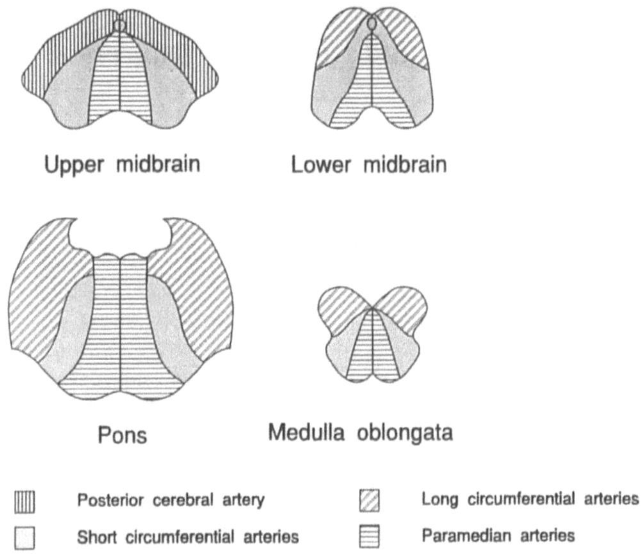

Fig. 1. Arterial territories of the brain-stem

Brain-stem Infarcts

Vascular Anatomy

The centromedian area of the brain stem is supplied by paramedian branches arising from the bifurcation of the basilar artery in the rostral midbrain, from the basilar artery in the lower midbrain and pons, and from the vertebral artery in the medulla oblongata (Fig. 1). The intermediolateral area depends on short circumferential branches arising from the posterior cerebral artery in the upper midbrain, on the basilar artery in the lower midbrain and pons, and on the vertebral artery in the medulla oblongata. The dorsolateral area is supplied by long circumferential branches arising from the superior cerebellar artery in the midbrain, from the anterior inferior cerebellar artery in the pons, and from the posterior inferior cerebellar artery in the medulla oblongata.

Results

Seventeen infarcts were located in the paramedian territory, two in the midbrain, and 15 in the pons (13 pontine infarcts were unilateral and two were bilateral). Five infarcts were located in the lateral region of the medulla oblongata; four were far lateral and one was more medially located (Table 1). Midbrain infarcts were characterized by ipsilateral third-nerve palsy and

Table 1. MRI topographic distribution of brain-stem infarcts (n = 22)

Localization		No.
– Midbrain	(paramedian arteries)	2
– Pons	(paramedian arteries)	15
– Medulla oblongata	(short circumferential arteries)	5

Table 2. Potential causes of brain-stem infarcts

Localization	Cause	n
– Paramedian midbrain infarcts	Cardiac	2/2
– Paramedian pontine infarcts	Cardiac	2/15
	Noncardiac	13/15
– Laterobulbar infarcts	Cardiac	2/5
	Noncardiac	3/5

contralateral hemiparesis. Paramedian pontine infarcts produced a variety of clinical syndromes. Ipsilateral sixth- and seventh-nerve palsies were associated with contralateral hemiparesis and hemiataxia in four patients. One patient had an ipsilateral sixth-nerve palsy with hemiataxia on the same side. One patient had an ipsilateral hemiataxia with contralateral hemiparesis. A contralateral hemiataxia was present in four patients and a pure contralateral hemiparesis in three. Both patients with bilateral paramedian infarcts had a pseudobulbar palsy, with hemiataxia in one case and hemiparesis in the other.

Four patients with an infarct in the lateral part of the medulla oblongata had ipsilateral ataxia and contralateral loss of pain and temperature sensation. In the patient whose bulbar infarct was more medially located, pain and temperature sensation loss involved also the face. Ipsilateral ninth-and tenth-nerve palsy was present in two patients, as was trigeminal hypoalgesia. Horner's syndrome was observed in three patients. In a very far laterally located infarct, clinical signs included only spontaneous nystagmus and ipsilateral gaze deviation. Table 2 presents the different causes of brain-stem infarcts.

Cerebellar Infarcts

Vascular Anatomy

In the cerebellum, the superior half of the hemisphere and vermis as well as the major part of the cerebellar white matter including the dentate nucleus are supplied (Fig. 2) by the superior cerebellar artery (SCA). The dorso-

territory infarcts are characterized by acute onset and pseudovestibular symptoms. lPICA territory infarcts, as well as mSCA and lSCA infarcts, present identically an ipsilateral limb ataxia and axial lateropulsion, but dysarthria seems to be a distinguishing sign of SCA territory infarcts (five of six patients in our series).

MRI gives useful information about the pathogenesis of infarcts. A noncardiac cause was found in 16 of 22 brain-stem infarcts. Practically speaking, cardiac embolism is rare in paramedian brainstem infarcts. In contrast, a cardioembolic origin has to be considered in every case of cerebellar infarction, particularly when the infarct is located medially, as was the case in nine of 15 patients in our series.

References

1. Amarenco P (1991) The spectrum of cerebellar infarctions. Neurology 41:973–979
2. Amarenco P, Hauw JJ (1989) Anatomie des artères cérébelleuses. Rev Neurol (Paris) 145:267–276
3. Amarenco P, Roulet E, Hommel M, Chaine P, Marteau R (1990) Infarction in the territory of the medial branch of the posterior inferior cerebellar artery. J Neurol Neurosurg Psychiatry 53:731–735
4. Bogousslavsky J, Meienberg O (1987) Eye-movement disorders in brainstem and cerebellar stroke. Arch Neurol 44:141–148
5. Duvernoy HM (1978) Human brainstem vessels. Springer, Berlin Heidelberg New York
6. Hommel M, Pollak P, Gaio JM (1988) Imagerie par résonnance magnétique et infarctus latérobulbaire. Rev Neurol (Paris) 144:272–278
7. Masson C, Chéron F (1990) Infarct in the territory of the medial branch of the PICA. J Neurol Neurosurg Psychiatry 53:1104–1105
8. Regli L, Regli F, Maeder P, Bogousslavsky J (1993) Magnetic resonance imaging using gadolinium contrast agent in lacunar brain infarcts. Arch Neurol 50:175–180
9. Savoiardo M, Bracchi M, Passerini A, Visciani A (1987) The vascular territories in the cerebellum and brainstem: CT and MR study. AJNR 8:199–209

Multifocal Ischemic Brain-Stem Lesions

B. Tettenborn

Neurologische Klinik, Universität Mainz, Langenbeckstraße 1,
55131 Mainz, Germany

Introduction

The clinical symptoms and signs of brain-stem ischemia are far less well defined than in ischemia due to carotid or middle cerebral artery disease. They consist of many different syndromes, and it is often difficult to correlate the clinical findings with a single functional or morphological lesion within the brain-stem.

Clinicotopographic correlation of small brain-stem infarcts was difficult to establish in the past because of the limitations of cranial computed tomography (CCT). Magnetic resonance imaging (MRI) now enables more precise topographic analysis and might supplement information from clinical examination [3]. Very small or only functional lesions might not even be detected by MRI, but electrophysiological studies can give additional information in correlation with the clinical symptoms and signs. The efficacy of electrophysiological studies in the evaluation of brain-stem involvement in demyelinating disease is well established, but their role in ischemic brain-stem disease has only started to become better clarified [32].

The combination of clinical findings and results from different electrophysiological studies and imaging techniques can give important information regarding the clinicotopographic correlation of brain-stem infarcts. Vascular studies, including ultrasound methods and angiography, allow the underlying vascular disease to be established.

If a patient has clinical symptoms and signs considered to be due to the sudden onset of brain-stem ischemia, we usually try to correlate the clinical findings to a single lesion. However, even within the anatomically very complex brain-stem, a unifocal lesion might not be able to explain all symptoms. Instead, separate small lesions within the brain-stem might be the cause of the complex clinical picture. One has to consider whether all clinical findings can be explained by a lesion within the vascular territory of one artery or one branch of an artery supplying the brain-stem, i.e., the vertebral arteries, the basilar artery, and the long circumferential arteries (posterior inferior cerebellar artery, anterior inferior cerebellar artery, superior cerebellar artery) or their branches. Multifocal lesions can occur in several branches of one artery or in different arteries.

L.R. Caplan and H.C. Hopf (Eds.)
Brain-Stem Localization and Function

Patients and Methods

Within a 2-year period we investigated 238 consecutive inpatients with clinical brain-stem symptoms of acute onset who had no major disturbance of consciousness thought to be due to ischemia. Electrophysiological, vascular, and imaging studies were performed in all patients and correlated with the clinical findings.

Electrophysiological Studies

Auditory Evoked Potentials. Short-latency, auditory evoked potentials (AEP) are a series of positive and negative waves recorded within the first 10 ms after an abrupt auditory stimulus. Data about the generator sites have been derived from intracranial recordings within or near the subcortical auditory pathways and from an analysis of the effects of lesions in humans as well as animals [5, 6, 9, 22, 23, 24, 34]. The generator site of wave I of the AEP is the eighth nerve. The origin of wave II is controversial; possible generators are the proximal part of the eighth nerve or the intra-axial infranuclear cochlear nerve [12]. Data from studies on patients with defined brain-stem lesions localize the major generators of wave III to the superior olivary complex in the caudal pontine tegmentum. Waves IV and V are generated in the lateral lemniscus within the upper pons and the inferior colliculus within the lower midbrain, respectively. In view of this anatomic basis, AEP abnormalities are expected in patients with pontine and mesencephalic lesions.

Electronystagmography. Electronystagmography (ENG) is used for monitoring eye movements. The potential between the cornea (the positive pole) and the retina (the negative pole) is measured (corneoretinal potential). Gaze is investigated for the presence of spontaneous, gaze-provoked, or positional nystagmus. Tests for vestibular ocular reflex function using caloric testing and visual ocular control regarding saccades and pursuit eye movements are performed. Different structures in the brain-stem and cerebellum are part of the oculomotor system [2, 10, 31]. Certain eye movements and ENG findings permit discrimination between a peripheral lesion within the labyrinth or eighth nerve and a central nervous system lesion, which provides additional information about localization within the brain-stem and cerebellum.

Blink Reflex. Electrical stimulation of the supraorbital nerve elicits two separate responses of the orbicularis oculi muscle: the early ipsilateral R1 component via a pontine pathway and the late ipsi- and contralateral R2 and R2c components relayed through a more complex route involving the pons and lateral medulla. These anatomic correlations originate from studies in patients with localized brain-stem lesions and from animal experiments [8, 17, 27]. Blink reflex latencies reflect conduction along the entire length of the supraorbital and facial nerves: therefore, pathological alterations of latencies do not necessarily indicate a pathological process within the brainstem. Afferent and efferent types of reflex abnormalities can be differentiated by analysis of the R2 component. Several studies suggest that supratentorial and mesencephalic structures also influence the R2 component [1, 18, 28]. Abnormalities of R2 and R2c can only be interpreted as a sign of a medullary lesion if a hemispheric or mesencephalic process is excluded. Isolated R1 changes can be due to pontine lesions [16, 19, 20, 30].

Masseter Reflex. The masseter or jaw reflex is elicited by a mechanical tap to the patient's jaw with a triggered hammer. The reflex is mediated by trigeminal fibers ascending to the mesencephalic nucleus and descending to the pontine motor nucleus [13, 14, 21, 25, 26]. Electrophysiologic evaluation depends on absolute normal values as well as on side-to-side comparison of the reflex responses recorded simultaneously from both masseter muscles. Differences in latencies of more than 0.5 ms and in amplitude of more than 50% indicate unilateral abnormality of the reflex arc, suggesting a pontomesencephalic lesion [13, 14, 33].

Vascular Studies

The invention of transcranial Doppler sonography (TCD) made it possible to evaluate the intracranial part of the vertebral artery and the basilar artery by suboccipital and transtemporal approach using a pulsed Doppler system operating at 2 MHz emitting frequency. The vertebral arteries can be insonated from their extradural to their intradural parts advancing to the junction of the vertebral arteries and along the basilar artery. The experience with TCD in vertebrobasilar large artery disease is still limited [4, 11, 29]. Cerebral angiography was performed if atherosclerotic disease in the vertebrobasilar arteries was suspected on the basis of clinical or TCD findings.

Imaging Techniques

All patients were investigated using CTT or MRI resonance imaging.

Other Studies

All patients underwent extensive laboratory studies and had thorough cardiac work-up.

Table 1. Multifocal infarctions

Type of infarction	Number of patients (n = 58)
Medulla and pons	7
Medulla and mesencephalon	3
Bilateral pons	10
Pons and mesencephalon	19
Bilateral pons and bilateral mesencephalon	6
Bilateral mesencephalon	5
Multiple lesions	8

Results

Among 238 patients with clinical brain-stem symptoms, the diagnostic methods, including clinical examination, electrophysiological studies, and imaging techniques, revealed multifocal brain-stem lesions in 58 patients. The distribution of the lesions within the brain-stem is shown in Table 1.

Clinical signs considered to be of good value in localization were specific cranial nerve palsies, crossed syndromes, vertical gaze palsy, and the one-and-a-half syndrome. These signs allowed localization rostrocaudally to medulla, pons, or midbrain and to medial or lateral tegmental regions. Clinical findings with little localizing value were other eye movement disturbances including nystagmus and horizontal gaze palsy, Horner's syndrome, hearing disturbance, and dysarthria. Other symptoms such as vertigo or nausea, gait abnormalities, hemiparesis, hemiataxia, or hemisensory signs proved to be of very little or no help in differentiating between medullary, pontine, and mesencephalic lesion within the brain-stem. Electrophysiological studies were only regarded as abnormal if there were definite pathological values. Follow-up examinations were performed in all patients and often revealed deterioration or improvement parallel to the clinical course.

Based on clinical findings alone, multifocal brain-stem abnormalities were found in 30 patients. Imaging techniques showed multifocal lesions in only 28 patients, whereas combined electrophysiological studies were able to show multifocal abnormalities in 37 patients. On MRI, lesions were only described as multifocal if there were several distinct signal changes on both sides at one brain-stem level or at different levels of the brain-stem on one or both sides. In 23 patients, neuroimaging was normal despite definite clinical signs of brain-stem dysfunction, and in seven patients only one lesion was detected by MRI, as compared to several lesions using clinical findings and electrophysiological results. In the majority of patients with electrophysiological findings of multifocal brain-stem lesions, the interpretation was based on pathological masseter reflex combined with abnormalities on electronystagmography or blink reflex testing; auditory evoked potentials were abnormal in only a few patients. In all patients with pathological results, control examinations were done.

All three methods – clinical examination, electrophysiological studies, and imaging techniques – correlated regarding multifocal ischemic brain-stem lesions in 14 patients. In the other 44 patients, only one or two of the above methods showed definite multifocal abnormalities.

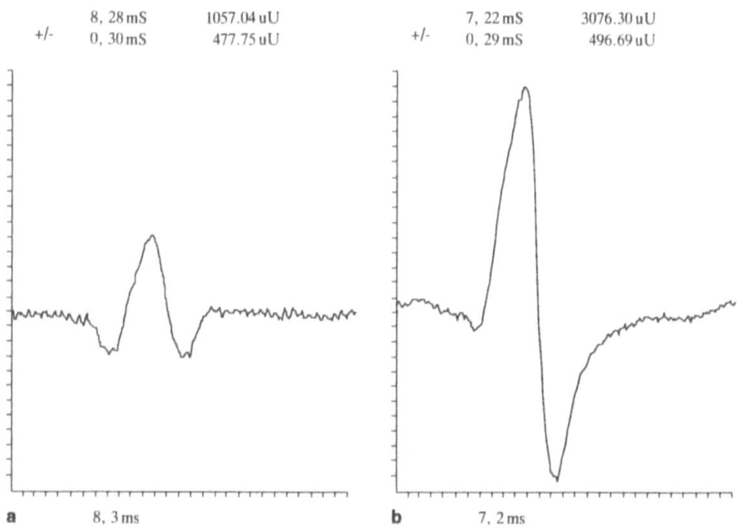

Fig. 1a,b. Masseter reflex of a patient with vertigo, nystagmus and, ataxia. **a** Examination 1 day after admission; pathological reflex on the left side. **b** Examination 6 months later; normalization of the masseter reflex

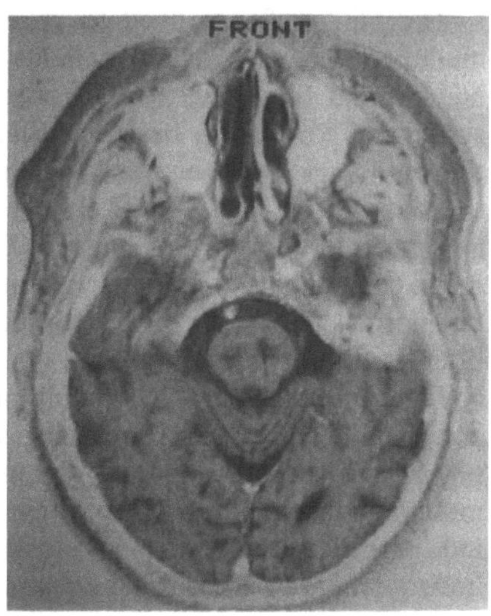

Fig. 2. Magnetic resonance imaging of a patient showing bilateral pontine lesions; left side of the brain showing on the right side

functional deficits without morphological changes. Whether the prognosis and therapeutic approach for multifocal ischemic brain-stem involvement should be different from those for unifocal lesions remains to be clarified.

References

1. Berardelli A, Accornero N, Cruccu G, Fabiano F, Guerrisi V, Manfredi M (1983) The orbicularis oculi response after hemispheral damage. J Neurol Neurosurg Psychiat 46:837–843
2. Bogousslavsky J, Fox AJ, Barnett HJM, Hachinski VC, Vinitski S, Carey LS (1986) Clinico-topographic correlation of small vertebrobasilar infarct using magnetic resonance imaging. Stroke 17:929–938
3. Bogousslavsky J, Meienberg O (1987) Eye-movement disorders in brainstem and cerebellar stroke. Arch Neurol 44:141–148
4. Büdinger HJ, Staudacher T (1987) Die Identifizierung der Arteria basilaris mit der transkraniellen Dopplersonographie. Ultraschall 8:95–101
5. Chiappa KH, Harrison JL, Brooks EB, Young RR (1980) Brainstem auditory evoked responses in 200 patients with multiple sclerosis. Ann Neurol 7:135–143
6. Curio G, Oppel F (1988) Intraparenchymateous ponto-mesencephalic field distribution of brain stem auditory evoked potentials in man. Electroencephalogr Clin Neurophysiol 69:259–265
7. Ferbert A, Buchner H (1991) Evozierte Potentiale in der Diagnostik ischämischer Hirnstammläsionen. Nervenarzt 62:460–466
8. Hacke W, Schaff C, Zeumer H (1983) Der Orbicularis-oculi-Reflex bei computertomographisch verifizierten Läsionen der hinteren Schädelgrube. Fortschr Neurol Psychiatr 51:313–324
9. Hashimoto I, Ishiyama Y, Yoshimoto S, Nemoto S (1981) Brainstem auditory evoked responses recorded directly from human brainstem and thalamus. Brain 104:841–859
10. Henn V, Büttner-Ennever JA, Hepp K (1982) The primate oculomotor system: I. Motoneurons II. Premotor system: a synthesis of anatomical, physiological, and clinical data. Hum Neurobiol 1:77–95
11. Hennerici M, Rautenberg W, Schwarz A (1987) Transcranial doppler ultrasound for the assessment of intracranial flow velovity, part 2. Surg Neurol 27:523–532
12. Hopf HC (1985) Die Generatoren der AEP-Wellen II–III. Aktuel Neurol 12:58–61
13. Hopf HC (1987) Vertigo and masseter paresis. A new local brainstem syndrome probably of vascular origin. J Neurol 235:42–45
14. Hopf HC, Gutmann L (1990) Diabetic 3rd nerve palsy: evidence for a mesencephalic lesion. Neurology 40:1041–1045
15. Hopf HC (1993) Topodiagnositic value of brain stem reflexes. Muscle & Nerve (in press)
16. Kayamori R, Dickins QS, Yamada T, Kimura J (1984) Brainstem auditory evoked potential and blink reflex in multiple sclerosis. Neurology 34:1318–1323
17. Kimura J (1975) Electrically elicited blink reflex in diagnosis of multiple sclerosis. Review of 260 patients over a seven year period. Brain 98:413–426
18. Kimura J, Wilkinson JT, Damasio H, Adams HR Jr, Shivapour E, Yamada T (1985) Blink reflex in patients with hemispheric vascular accident (CVA). J Neurol Sci 67:15–28
19. Koshbin S, Hallett M (1981) Multimodality evoked potentials and blink reflex in multiple sclerosis. Neurology 31:138–144
20. Krämer G, Lowitzsch K, Besser R, Hopf HC (1986) Isolierte R1-Veränderungen des Blinkreflexes. In: Lowitzsch K (ed) Hirnstammreflexe. Thieme, Stuttgart, pp 124–131
21. Kubota K, Narita N, Ohkubo K, Hosaka K, Nagae K, Lee MS, Kawamoto T, Kubota M, Odagiri N (1988) Central projection of proprioceptive afferents arising from maxillo-facial regions in some animals by HRP-labeling technique. Anat Anz 165:229–251

22. Legatt AD, Arezzo JC, Vaughan HG Jr (1986) Short-latency auditory evoked potentials in the monkey: II. Intracranial generators. Electroencephalogr Clin Neurophysiol 64:53–73
23. Markand ON, Farlow MR, Stevens JC, Edwards MK (1989) Brain-stem auditory evoked potential abnormalities with unilateral brain-stem lesions demonstrated by magnetic resonance imaging. Arch Neurol 46:295–299
24. Moller AR, Janetta PJ (1983) Interpretation of brainstem auditory evoked potentials: results from intracranial recordings in humans. Scand Audiol 12:125–133
25. Nomura S, Mizuno N (1985) Differential distribution of cell bodies and central axons of mesencephalic trigeminal nucleus neurons supplying the jaw-closing muscles and periodontal tissue. Brain Res 359:311–319
26. Ongerboer de Visser BW, Goor C (1976) Jaw reflexes and masseter electromyograms in mesencephalic and pontine lesions: an electrodiagnostic study. J Neurol Neurosurg Psychiatry 39:90–92
27. Ongerboer de Visser BW, Kuypers HGJM (1978) Late blink reflex changes in lateral medullary lesions. Brain 101:285–294
28. Ongerboer de Visser BW (1983) Comparative study of corneal and blink reflex latencies in patients with segmental or with cerebral lesions. In: Desmedt JE (ed) Motor control mechanisms in health and disease. New York, Raven Press, pp 757–772
29. Ringelstein EB, Zeumer H, Poek K (1985) Non-invasive diagnosis of intracranial lesions in the vertebrobasilar system. A comparison of doppler sonographic and angiographic findings. Stroke 16:848–855
30. Sanders EACM, Ongerboer de Visser BW, Barendswaard EC, Arts RJHM (1985) Jaw, blink, and corneal reflex latencies in multiple sclerosis. J Neurol Neurosurg Psychiat 48:1284–1289
31. Spector RH, Troost BT (1981) The ocular motor system. Ann Neurol 9:517–525
32. Tettenborn B, Caplan LR, Krämer G, Hopf HC (1992) Electrophysiology in ischemic brain stem disease. In: Berguer R, Caplan LR (eds) Vertebrobasilar arterial disease. Quality Medical Publishing, St. Louis, pp 124–129
33. Yates SK, Brown WF (1981) The human jaw jerk: electrophysiologic methods to measure the latency, normal values, and changes in multiple sclerosis. Neurology 31:632–634
34. York DH (1985) Correlation between a unilateral midbrain-pontine lesion and abnormalities of brain-stem auditory evoked potential. Electroencephalogr Clin Neurophysiol 65:282–288

Craniocaudal Disintegration of Neurophysiological Findings in Progressive Coma

A. Ferbert

Klinik für Neurologie, Mönchebergstraße 41-43, 34125 Kassel, Germany

Introduction

Since the early days of this century, it has been recognized that space-occupying lesions of the brain lead to displacement of neighboring structures. In 1909, Cushing [9] described herniation of the cerebellar tonsils into the foramen magnum, whereas the importance of transtentorial herniation was not appreciated until the early 1930s. The sequelae of the latter have been extensively studied: The uncus of the temporal lobe is pushed into the tentorial notch. Still later, the mechanism of cingular herniation with displacement of the cingular gyrus under the falx cerebri was described. There is craniocaudal displacement of the midbrain, leading to pressure on the third cranial nerve and occasionally on the posterior cerebral artery at the incisura tentorii. Later, the pons and medulla oblongata and, to a lesser degree, the basilar artery with its branches to the pons are also displaced downwards. This gives rise to severe tissue damage in the brain stem, caused by the pressure itself or by secondary ischemia.

Given the high sensitivity of electrophysiological methods in patients with primary brain-stem lesions, we might also expect the same sensitivity for secondary brain-stem lesions. Those electrophysiological methods that test pathways running through the brain stem such as brain-stem auditory evoked potentials (BAEP) and somatosensory evoked potentials (SEP) are capable of demonstrating the level at which the brain stem is injured by the herniation.

This review will focus on evoked potentials. However, other methods for evaluation of herniation have been used: Assessment of the level of coma can also be performed by means of EEG [e.g. 35], the disadvantage being that the EEG is modified by sedative drugs. Visual evoked potentials usually deteriorate early in the course of herniation. They may, however, be preserved in primary infratentorial lesions even if all brain-stem functions are lost [11]. Brain-stem reflexes that are of great value in primary brain-stem lesions have found less broad application in secondary brain-stem lesions, as reflex excitability is considerably dependent on the function of the reticular formation. Abnormal findings may then be attributed to a lesion of the

L.R. Caplan and H.C. Hopf (Eds.)
Brain-Stem Localization and Function

reflex arc itself or, alternatively, to abnormal influences from the reticular formation, particularly if they are symmetrical.

Clinical Features

One of the first signs of herniation is neck rigidity, but this sign is not always present [34]. Certainly, one of the most significant signs is progressive loss of consciousness, from clouding to deep coma. The oculomotor nerve is stretched and often gives rise to a dilated, nonreacting pupil, mostly on the side ipsilateral to the space-occupying lesion. Further signs of a third-nerve lesion follow later. Papilledema will be found only rarely in cases with acute herniation and without pre-existing elevated intracranial pressure. Some authors [27, 28] have differentiated central herniation with early loss of consciousness from lateral herniation. In the latter, the uncus is displaced into the tentorial notch and, according to these authors, oculomotor nerve signs may precede clouding of consciousness. These authors [27, 28] described clinical features of what they called the early and late diencephalic stages, the midbrain/upper pons stage, the lower pons/medulla stage, and the medullary stage.

A similar way of correlating clinical signs with particular anatomical structures [18] was to describe the clinical features of four different stages of an "acute midbrain syndrome" as a result of tentorial herniation. The signs ranged from clouding of consciousness with normal reaction to painful stimuli, reacting pupils, and roving eye movements in the first stage to coma, extensor rigidity, hyperreflexia, tachycardia, hypothermia, dilated, sluggishly reacting pupils, and lack of spontaneous eye movements but preserved corneal reflex in the fourth stage of the midbrain syndrome. The authors [18] further attributed a "transitional stage into the acute bulbar syndrome" to foraminal herniation with loss of all brain-stem reflexes and lack of posturing with preserved spontaneous respiration that finally leads to loss of all brain-stem functions in the brain death syndrome. This sequence of signs has many individual variations, particularly with respect to the temporal course. It is, however, important to note that this sequence applies only to craniocaudal herniation and cannot be found in coma due to primary brain-stem lesions or intoxication.

Brain-stem Auditory Evoked Potentials (BAEP)

As different waves of the BAEP are generated at different levels of the brain stem, they are particularly suited to define the level of a brain-stem lesion. The craniocaudal loss of brain-stem functions is matched by a progressive loss of BAEP waves, beginning with waves of longer latency generated in more cranial structures and advancing to earlier waves generated in the lower brain stem. Despite the large amount of literature on the impact of BAEP in predicting the outcome of secondary brain-stem compression, e.g., in brain injury [1, 8, 19, 30] and in the diagnosis of brain death [4, 10, 23, 26, 32], there is only a small amount of data precisely correlating different clinical stages as mentioned above with certain BAEP patterns. This is particularly true for earlier stages of herniation.

The lack of correlation might be due to the early use of artificial ventilation that has become common practice in recent years, often requiring sedation of the patient at least in the early stages of herniation, thus preventing the clinician from performing a proper neurological examination. Although the therapeutically induced barbiturate coma is not generally recommended, it is still used in some centers. BAEP are nearly completely unaffected by sedative drugs unless very high doses of barbiturates are used, leading to mild amplitude drop of BAEP. In contrast, the clinical examination (particularly consciousness, corneal reflex, reaction to painful stimuli) is highly sensitive to sedative or analgetic drugs. Therefore, BAEP can be used as objective determinants of brain-stem functions in herniation. In order to provide continuous monitoring [20], special computer programs have been developed [17].

Serial examinations of BAEP in a patient with herniation until brain death are shown in Fig. 1. The very first abnormality that occurs is a prolonged interpeak latency III–V with some amplitude reduction of wave V. This does not occur, however, in the very early stage of herniation, in which the BAEP may be normal. This can be explained by the generator of wave V: Whereas earlier work pointed to the midbrain, there is now some evidence that the pons generates wave V [12, 14]. Amplitude reduction or loss of wave V leaving wave IV intact leads to "pseudo" shortening of latency of the wave "IV/V" complex. Wave IV may sometimes mistakenly be regarded as wave V (Fig. 2). As the herniation progresses, waves V and IV are lost. Once waves IV and V are completely lost, changes of wave III soon follow. Changes of III–V occur in a relatively symmetrical fashion. Pronounced asymmetries of waves III–V between right and left ear stimulation are more likely to be produced by local brain-stem lesions [12, 14].

When all brain-stem functions including spontaneous respiration cease, then only wave I and occasionally wave II can be observed [4, 32] (see Fig. 1). These waves are generated by the eighth nerve and not by the brainstem and therefore do not contradict the diagnosis of brain death. The finding of an isolated wave I usually is rather stable, but wave I can also finally disappear, presumably due to decreased perfusion by the internal auditory artery (Fig. 3). An interesting phenomenon is the increase in amplitude of wave I that occasionally occurs before it disappears [5, 17, 32]. This may be explained by the loss of an inhibitory centrifugal tract to the cochlea. Alternatively, it could reflect the increased excitability that some neurons show under hypoxic conditions [21].

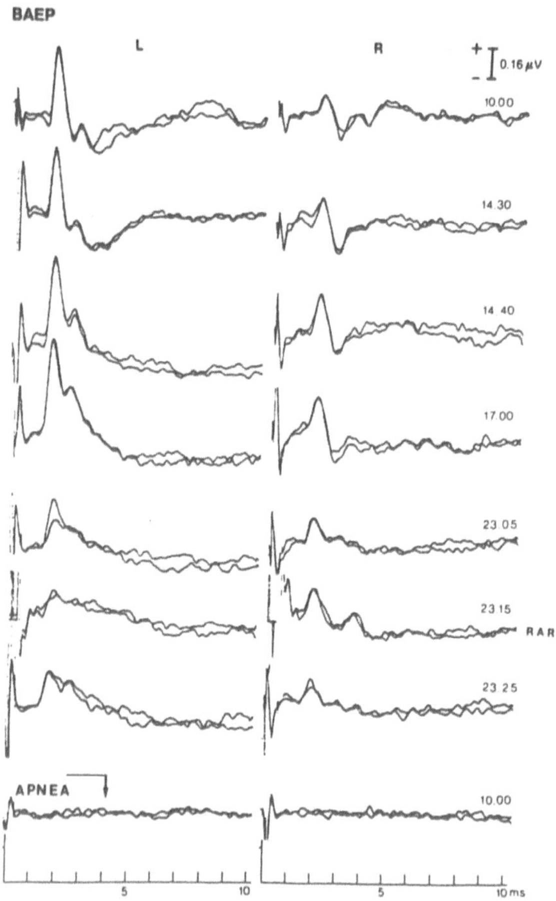

Fig. 3. Serial BAEP recordings of a patient with a space-occupying infarction of both the middle and anterior cerebral artery territories. From at least the *second trace* onwards, there is no wave preserved after wave II, despite preserved spontaneous respiration. Apnea occurred only 20 h later (*bottom trace*). While in infratentorial lesions such an observation is not uncommon, this is the only reported case with such a constellation in a supratentorial lesion. [From 3]

Somatosensory Evoked Potentials (SEP)

There is an abundant amount of literature on the significance of cortical SEPs as a prognostic tool in patients with head injury. Even though head injury is not the particular topic of this paper, one can reasonably assume that the crucial factor in head injury is edema leading to herniation, and that electrophysiological data can be generalized to herniations of different causes. Several findings have been described that are of prognostic importance:

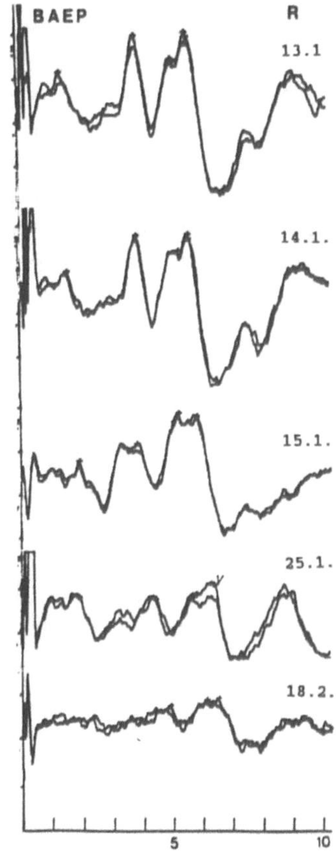

Fig. 4. Serial BAEP recordings of a patient with a small supratentorial hemorrhage. Prolonged intensive-care treatment was necessary because of pneumonia. The patient showed no signs of neurological deterioration. Progressive loss of BAEP waves has to be explained as the consequence of ototoxic factors and not as progressive brain-stem dysfunction

1. Absence of cortical SEP bilaterally has been found to be an indicator of poor prognosis [8, 29].
2. The absolute latency of the N20 component is normal or only marginally prolonged. However, by considering the central somatosensory conduction time (CSCT), i.e., the latency difference between the N13 cervical response and the N20 component, one can derive prognostic information [22, 31]. Most work concerning CSCT has been done with a reference C7/Fz. However, the response obtained using this lead is not really an extracephalic reference potential [24, 25] and can itself be affected in foraminal herniation.

Although the absence of cortical SEP is of prognostic value, the localizing value of this finding concerning the exact level of herniation is poor. It can be found in the complete apallic syndrome [2], in the advanced midbrain syndrome, and in the bulbar syndrome. Can the level of craniocaudal disintegration be better determined by recording subcortical SEP?

Subcortical SEP consist of a series of low-amplitude potentials that can reliably be assessed only when a high signal-noise ratio can be achieved.

Their clear recognition can be hampered by the fact that they overlay each other. Loss of one of these waves may therefore be hidden by the persistence of a different wave of similar latency. Different recording montages can help to differentiate such waves of similar latency. Thus, for identification of a potential one should indicate not only polarity and time of occurrence but also the recording sites used.

The P14 component recorded between vertex and a noncephalic electrode is generated in the brain stem and is lost in brain-stem compression leading to brain death. Loss of P14 in this recording is particularly easy to recognize if it is distinct from P13. As those waves are merged in some healthy subjects, loss of P14 has then to be assumed when there is a significant loss of amplitude of this wave. Therefore, a reliable judgement about the presence or absence of P14 can best be made if a baseline recording is available for comparison. The N13 component in the recording C7 – jugular bone is a purely spinal potential and may therefore best be used as reference for arrival of the volley in the spinal cord. It is therefore unaffected in brain

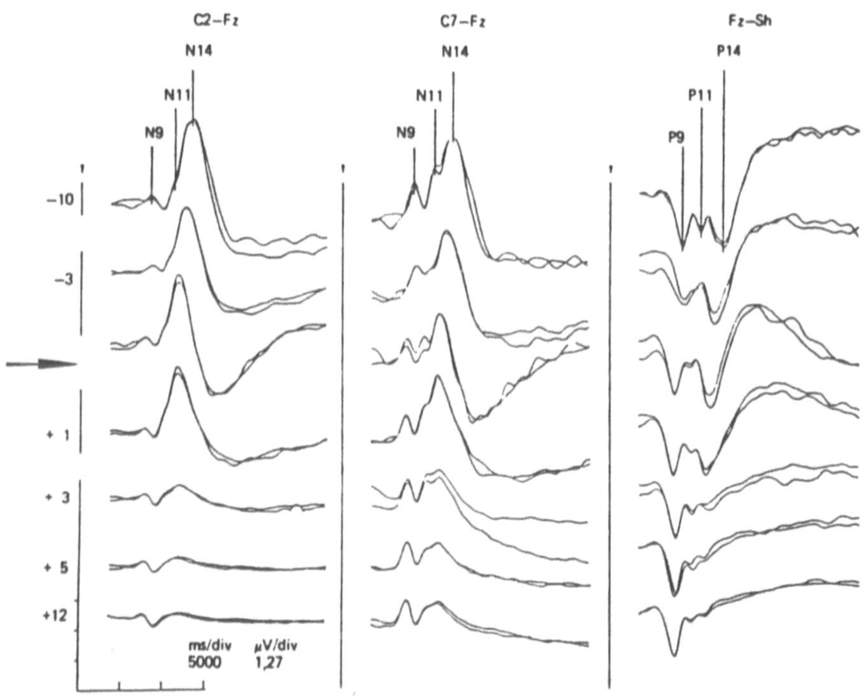

Fig. 5. Subcortical SEP before and after loss of clinical brain-stem functions that occurred at the time of *trace 3*. Figures on the left indicate time before and after the time of trace 3. The Fz-shoulder recording shows loss of P14. As N13/14 in the C7-Fz recording are not separated from each other, loss of P14 is reflected only as amplitude drop of the N13/14 component in the C7-Fz recording. Minor latency changes of earlier components could be explained by changes of body temperature. [From 7]

death unless there is severe spinal cord ischemia [6, 7]. The commonly used C7 Fz recording is somehow in between the above-mentioned recordings, in that N13/14 in herniation is less affected than the P14 (recorded Cz – noncephalic) but clearly more than the spinal potential N13 (C7 – jugular bone). For a more complete description of SEP components and their relation to the brain stem see Maugière's discussion in this volume.

Figures 5 and 6 show two examples of SEP monitoring in progressive herniation and in brain death. In Fig. 5, loss of the P14 component in the Fz-shoulder lead is matched by a significant amplitude drop of N13/14 in the C7/Fz lead, suggesting that the latter component consists of at least two generators. Their components occur about at the same time and therefore overlay each other in a single wave. Loss of the P14 component is clearly seen in this recording. The loss of P14 is less clear in Fig. 6, in which there is gradual attenuation of waves over several hours after loss of all brain-stem functions. The C7 – jugular bone recording shows the "true", spinal N13 that is unaffected by the herniation.

To summarize, craniocaudal herniation that has progressed to the level of the medulla oblongata leads to loss of P14, recorded best from Cz to a noncephalic reference electrode. However, clear separation of these com-

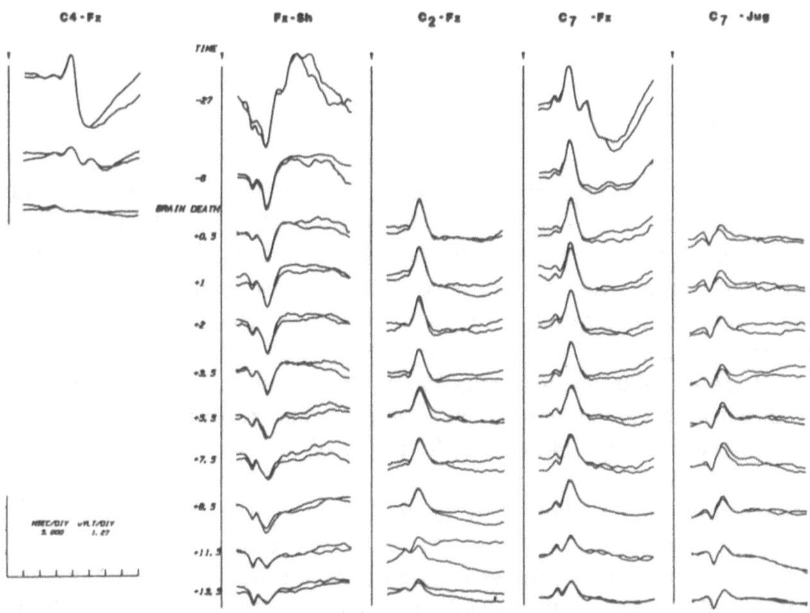

Fig. 6. Serial recording of cortical and subcortical SEP before and in brain death. Neither the Fz – shoulder nor the C2 or C7-Fz recording shows a clear loss of a component in brain death. This is because the subcortical SEP components were not well separated from each other at the beginning. Loss of P14 has to be assumed from the gradual amplitude drop of the responses. There is no attenuation of the N13 component in the C7 – jugular bone recording because in this lead only spinal activity is recorded. [From 6]

ponents is sometimes difficult, and the relative representation of the brain stem in the SEP is less than that in the BAEP.

Transcranial Magnetic Stimulation (TCMS)

The value of transcranial magnetic brain stimulation in coma has only recently been investigated in some centers. Theoretically, any bilateral disruption of the pyramidal tract – whether in the forebrain hemispheres, the pons, or in the upper spinal cord – will lead to abolition of motor evoked potentials. Thus, there is no response to stimulation techniques that would allow physicians to specifically investigate functions of the brain stem. It is possible, however, to test the integrity of the lower motor neurons by stimulating over the lower cervical cord and recording from small hand muscles.

In one study, 51 patients with brain death due to primary brain lesions were investigated by means of TCMS. In none of them were motor evoked potentials (MEP) recorded after cortical stimulation, whereas cervical stimulation evoked responses in all patients. In three patients with coma grade III, MEP were preserved while cortical SEP were already lost. This led the authors to conclude that MEP appeared more resistant than cortical SEP to increasing severity of coma in these three patients [15].

Conclusion

Several neurophysiological tools can be used to monitor brain-stem function in patients with craniocaudal herniation. BAEP and SEP have the advantage of being nearly unaffected by sedative drugs and are therefore particularly suited for this purpose. The best evaluation of the level of craniocaudal disintegration is provided by BAEP. However, the early stage of herniation may not be detected by BAEP. In the late stage of herniation BAEP are of value in the evaluation of brain death. SEP can also be used for monitoring caudal brain-stem function when recording P14 in a Cz – noncephalic montage. However, the utility of SEP depends on how clearly several subcortical SEP components are separated from each other in the preherniation stage. There may be patients in whom the presence or absence of P14 is difficult to assess. At the present time, transcranial magnetic stimulation adds little to the evalution of craniocaudal disintegration.

If we take these electrophysiological measures together with methods of direct recording of intracranial pressure, herniation can be reliably assessed. A problem that is insuffiently resolved, however, is the medical treatment of severe brain edema that causes such herniation.

References

1. Anderson DC, Bundlie S, Rockswold G (1984) Multimodality evoked potentials in closed head trauma. Arch Neurol 41:369–374
2. Biniek R, Ferbert A, Rimpel J, Paepke U, Berns TH, Schuchard V, Heitmann R (1989) The complete apallic syndrome – a case report. Intensive Care Med 15:212–215
3. Biniek R, Ferbert A, Buchner H, Brückmann H (1990) Loss of brainstem acoustic evoked potentials with spontaneous breathing in a patient with supratentorial lesions. Eur Neurol 30:38–41
4. Buchner A, Ferbert A, Brückmann H, Zeumer H, Hacke W (1986) Zur Validität der frühen akustisch evozierten Potentiale in der Diagnose des Hirntods. Z EEG EMG 17:117–122
5. Buchner H, Ferbert A, Scherg M, Hacke W (1986) Evoked potential monitoring in brain death – generators of BAEP and SEP. In: Kunze K, Zangemeister WH, Arlt A (eds) Chinical problems of brainstem disorders. Thieme, Stuttgart, pp 130–133
6. Buchner H, Ferbert A, Hacke W (1988) Serial recording of median nerve stimulated subcortical somatosensory evoked potentials (SEPs) in developing brain death. Electroencephalogr Clin Neurophysiol 69:14–23
7. Buchner H, Schildknecht M, Ferbert A (1991) Spinale und subkortikale somatosensibel evozierte Potentiale; Vergleich mit der Lokalisation spinaler, medullärer und pontiner Läsionen und im Hirntod. Z EEG EMG 22:51–61
8. Cant BR, Hume AL, Judson JA, Shaw NA (1986) The assessment of severe head injury by short-latency somatosensory and brain-stem auditory evoked potentials. Electroencephalogr Clin Neurophysiol 65:188–195
9. Cushing H (1909) Some aspects of the pathological physiology of intracranial tumors. M & SJ, Boston
10. Ferbert A, Riffel B, Buchner H, Ullrich A, Stöhr M (1985) Evozierte Potentiale in der neurologischen Intensivmedizin – eine Standortbestimmung. Aktuel Neurol 12:193–198
11. Ferbert A, Buchner H, Ringelstein EB, Hacke W (1986) Isolated brainstem death. Case report with demonstration of preserved visual evoked potentials (VEP). Electroencephalogr Clin Neurophysiol 65:157–160
12. Ferbert A, Buchner H, Brückmann H, Zeumer H, Hacke W (1988) Evoked potentials in basilar artery thrombosis: correlation with clinical and angiographic findings. Electroencephalogr Clin Neurophysiol 69:136–147
13. Ferbert A, Buchner H, Ringelstein E, Hacke W (1989) Brain death from infratentorial lesion; clinical, neurophysiological and transcranial Doppler ultrasound findings. Neurosurg Rev 12 Suppl 1:340–347
14. Ferbert A, Buchner H, Brückmann H (1990) Brainstem auditory evoked potentials and somatosensory evoked potentials in pontine haemorrhage. Brain 113:49–63
15. Firsching R, Wilhelms S, Csescei G (1992) Pyramidal tract function during onset of brain death. Electroencephalogr Clin Neurophysiol (in press)
16. Gail S, Ferbert A, Behrendt W (1991) Beeinflussen intensivmedizinische Maßnahmen die frühen akustisch evozierten Hirnstammpotentiale? Intensivmedizin 28:396–397
17. Garcia-Larrea L, Bertrand O, Artru F, Pernier J, Mauguiere F (1987) Brain-stem monitoring. Preterminal BAEP changes observed until brain death in deeply comatose patients. Electroencephalogr Clin Neurophysiol 68:446–457
18. Gerstenbrand F, Lücking CH (1970) Die akuten traumatischen Hirnstammschäden. Arch Psychiatr Nervenkr 213:264–281
19. Greenberg R, Newlon P, Hyatt M, Narayan R, Becker D (1981) Prognostic implications of early multimodality evoked potentials in severely head injured patients. J Neurosurg 55:227–236
20. Hacke W (1985) Neuromonitoring. J Neurol 232:125–133
21. Heiss WD, Hayakawa T, Waltz AG (1976) Cortical neuronal function during ischemia. Arch Neurol 33:813–820

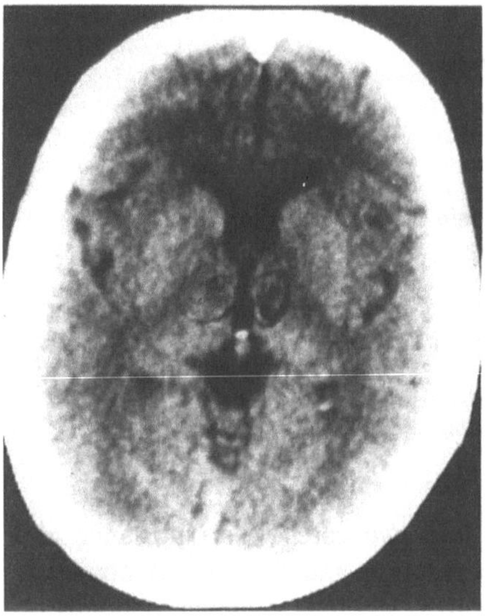

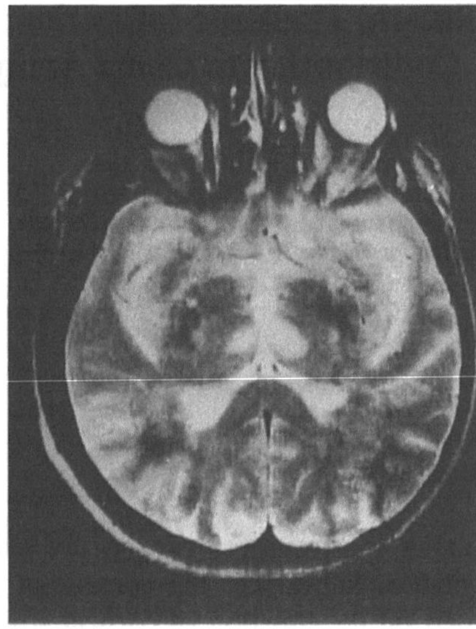

Fig. 1 (*left*). CT of patient 2, showing bilateral thalamic infarction

Fig. 2 (*right*). MRI of patient 3, showing bilateral thalamic infarction

day, CT showed hypodense areas in the medial thalamus bilaterally (Fig. 1). All other investigations were normal.

3. On admission this 57-year-old woman was somnolent with no signs of meningism. There were no focal abnormalities except for a loss of the vertical oculocephalic reflex. EEG revealed coma 1, in the classification of Fischgold. Though awake on the next day, the patient had very poor memory and was disoriented. The neurological examination revealed the rare sign of an isolated vertical downward gaze paralysis. On day 2 CT and MRI showed a bilateral medial thalamic infarction (Fig. 2). All other investigations were normal. After 3 years, the neuropsychological symptoms had disappeared. Only a slowing of downward saccades remained.
4. After exhausting work with a personal computer, a 33-year-old man noticed double vision during a phone call. He arrived at the hospital unconscious and was somnolent after 6 h. The right eye was in abduction but horizontal pursuit movements were otherwise normal. There was a complete downward and only a discrete upward vertical gaze paralysis. The neurological status was otherwise normal. On day 3 no lesion was detected on the CT but the MRI showed a bilateral medial thalamic infarction, which was larger on the right side than on the left. All further findings were normal. After 2 weeks there was only mild impairment of all vertical eye movements.

Discussion

Different patterns of thalamic infarctions have been described in careful studies [1, 2]. The aim of this paper is to call attention to lesions in the mesodiencephalic area and medial thalamus as a potential cause of acute impairment of vigilance. Vigilance is regulated by different neuronal systems, all originating in the reticular nuclei of the brain stem. They are characterizied by specific functions and transmitters. The best known of these systems is the dorsal root of the cholinergic ascending reticular activating system (ARAS) that projects from the reticular formation to the parafascicular and centromedian nuclei of the medial thalamus and from there to the neocortex [3, 4]. Stimulation of this system in animals induces the arousal reaction. Further activating systems are the ventral cholinergic and the noradrenergic systems [5]. Serotoninergic neurons seem to have deactivating functions.

It is important to note that only damage to both sides of the vigilance-regulating systems causes unconsciousness. Due to the close topographical vicinity in the mesodiencephalic region, however, even small lesions can alter the systems bilaterally. For clinical examination in mesodiencaphalic lesions it is important to carefully analyze the vertical oculomotor system. Lesion [6] as well as stimulation [7, 8] experiments have demonstrated that the upward eye movements are based on the function of the pretectal area and its connections through the posterior commissure. Bilateral lesion of this area or unilateral lesions involving the posterior commissure lead to upward gaze paralysis. An isolated downward gaze paralysis is extremely rare and can be observed only with two distinct small lesions in the prerubral fields. Due to the close vicinity of the pretectal and prerubral areas, the most common syndrome is complete vertical gaze paralysis. The described structures for the regulation of vertical eye movements are located close to the ARAS. This area is often supplied by the unpair thalamicsubthalamic artery [9, 10].

All patients in our study presented with impaired consciousness, ranging from somnolence to coma, as the main initial syndrome. In the absence of increased intracranial pressure, metabolic or inflammatory diseases as well as intoxication were initially considered. In only one patient was the correct diagnosis immediately given by the detection of a dissociation of the vestibulo-ocular reflex, which was preserved in the horizontal plane but abolished vertically.

There is extensive documentation of similar cases in the literature [11, 12]. Various other signs were also reported in cases of bilateral thalamic infarction, such as palilalia [13], oral dyskinesia [14], and loss of psychic self-activation [15]. An embolic etiology was documented for some patients with vertical gaze paralysis and unconsciousness in another study [1]. We were not able to detect the pathogenesis or etiology of the vascular lesions in our patients despite intensive cardiovascular investigations including transthoracic and transesophageal echocardiography as well as doppler sonography of the cervical arteries.

Generally, the prognosis was good in all cases. After several weeks only descrete amnestic or oculomotor syndromes persisted.

References

1. Graff-Radford NR, Damasio H, Yamada T, Eslinger PJ, Damasio AR (1985) Nonhaemorrhagic thalamic infarction. Clinical, neurophysiological and electrophysiological findings in four anatomical groups defined by computerized tomography. Brain 108:485–516
2. Bogousslavsky J, Regli F, Uske A (1988) Thalamic infarcts: clinical syndromes, etiology, and prognosis. Neurology 38:837–848
3. Moruzzi G, Magoun HW (1949) Brain-stem reticular formation and activation of the EEG. Electroencephalogr Clin Neurophysiol 1:455–473
4. Von Cramon D (1984) Die neuronalen und humoralen Voraussetzungen des Bewußtseins. In: de Pai AW, Dageförde J, Neundörfer B, Scriba PC (eds) Die unklare Bewußtlosigkeit. Zuckschwerdt, Münich, pp 3–12
5. Koella D (1982) Vigilance – a concept and its neurophysiological and biochemical implications. In: Passouant, Oswald (eds) Pharmacology of the status of alertness. Pergamon, Oxford, pp 171–178
6. Büttner-Ennever JA, Büttner U, Cohen B, Baumgartner G (1982) Vertical gaze paralysis and the rostral interstitial nucleus of the medial fasciculus. Brain 105:125–149
7. Kömpf D, Pasik T, Bender MB (1979) Downward gaze paralysis in monkeys. Stimulation and lesions studies. Brain 102:527–558
8. Henn U, Lang W, Hepp K, Reisine H (1984) Experimental gaze palsies in monkeys and their relations to human pathology. Brain 107:619–639
9. Percheron G (1976) Les artères du thalamus humain. II. Artères et territoires thalamiques paramédines de l'artère basilaire communicante. Rev Neurol (Paris) 132:309–324
10. Segarra JM (1970) Cerebral vascular disease and behaviour. I. The syndrome of the mesencephalic artery (basilar artery bifurcation). Arch Neurol 22:408–418
11. Castainge P, Lhermitte F, Buge A, Escourolle R, Hauw JJ, Lyon-Caen O (1981) Paramedian thalamic and midbrain infarcts: clinical and neuropathological study. Ann Neurol 10:127–148
12. Gentilini M, de Renzi E, Crisi G (1987) Bilateral paramedian thalamic artery infarcts: report of eight cases. J Neurol Neurosurg Psychiatry 50:900–909
13. Yasuda Y, Akiguchi I, Ino M, Nabatabe H, Kameyama M (1990) Paramedian thalamic and midbrain infarcts associated with palilalia. J Neurol Neurosug Psychiatry 53:797–799
14. Combarros O, Gutiérrez A, Pascal J, Berciano J (1990) Oral dyskinesias associated with bilateral thalamocapsular infarction. J Neurol Neurosurg Psychiatry 53:168–169
15. Bogousslavsky J, Regli F, Delalaye B, Delalaye-Bischoff A, Assal G, Uske A (1991) Loss of psychic selfactivation with bithalamic infarction. Acta Neurol Scand 83: 309–316

Cerebellar System, Vestibular Function, FAEP

Topical Signs of Cerebellar Disease

P. Thier and J. Dichgans

Neurologische Klinik, Kliniken Schnarrenberg, Eberhard Karls-Universität, Hoppe-Seyler-Straße 3, 72076 Tübingen, Germany

Since the pioneering studies of Gordon Holmes [25], numerous reports have tried to relate functional deficits in patients with cerebellar lesions to the site of the lesion. While there is continuing uncertainty as to the functional role of the cerebellum, previous work has been able to derive a fairly stable concept of the topography of the functional deficits resulting from cerebellar lesions. Since a number of comprehensive reviews on the subject are available [19, 26], we will content ourselves with a brief summary of the well-established oculomotor and motor features of cerebellar disease and touch briefly on recent thoughts about the relation of the cerebellum to higher brain functions. Special emphasis will be given to observations which challenge the traditional view of the oculomotor role of the posterior vermis – a change in perspective which we feel may have implications for the diagnosis of cerebellar disease. The emphasis on this specific subject is necessarily subjective and reflects the authors' own interests and activities.

Oculomotor Deficits Related to the "Vestibulocerebellum"

Two separate parts of the cerebellum, the posterior vermis with the fastigial nucleus on the one hand, and the flocculonodular lobe on the other hand, have been shown to play a role in the organization of eye movements. The flocculonodular lobe consists of the paired flocculi lying next to the eighth cranial nerve and the unpaired nodulus, a midline structure, located immediately below the inferior medullary velum. The flocculonodular lobe and the paired paraflocculi, located immediately adjacent to the flocculi, are dominated by input from the vestibular nuclei, to which they project back in turn [36, 37, 73]. This connectivity suggests a major role in the control and modification of the vestibulo-ocular (VOR) and optokinetic (OKR) reflexes mediated by the vestibular nuclei (see [71] for a detailed discussion). Along this line, a major part of the syndrome resulting from lesions of the flocculonodular lobe can be understood as release from signals modifying the VOR and the OKR. When discussing the consequences of floccular lesions one should take into consideration that the lesions reported in the literature

L.R. Caplan and H.C. Hopf (Eds.)
Brain-Stem Localization and Function

bances result. Such pursuit disturbances were first demonstrated by Zee et al. [77].

Figure 1A, taken from their original publication compares pursuit of a target moving predictably before and after bilateral flocculectomy. The outcome of this experiment was characterized by a reduction of pursuit gain

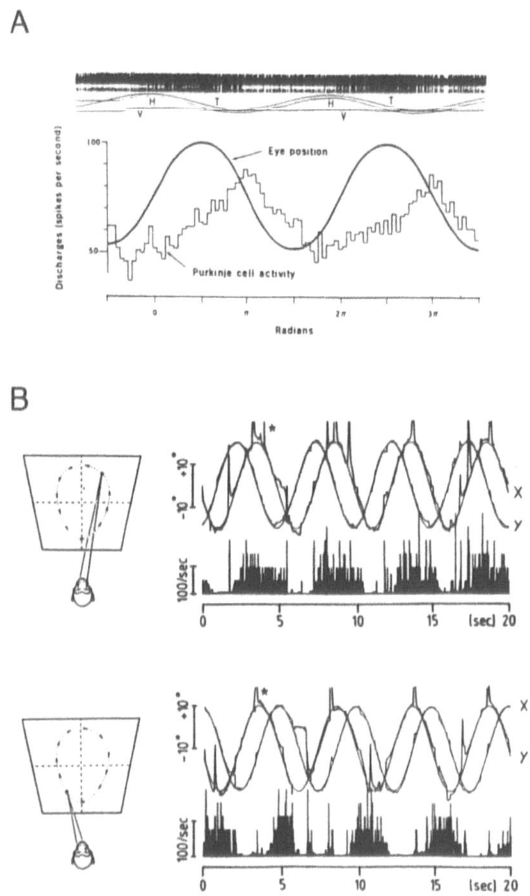

Fig. 2 A,B. Two examples of neurons activated by smooth-pursuit eye movements. The neuron shown in **A** was recorded from the posterior vermis, whereas the one shown in **B** was recorded from the dorsolateral pontine nucleus (DLPN), a major source of afferents integrated by the posterior vermis. **A** This response is presented in the form of an original spike record (*top*) and a poststimulus time histogram averaging the discharge rate over many cycles of sinusoidal, horizonal smooth pursuit (*bottom*). *H*, horizontal electro-oculogram; *V*, vertical electro-oculogram; *T*, target position. (Modified from [31]) **B** Smooth pursuit guided by a visual target moving in a counterclockwise (*top*) and clockwise manner (*bottom*) on a circle in the frontoparallel plane. *X* and *Y* are continuous records of the x and y components of eye and target movement, respectively. Discharge is plotted in the form of the momentary discharge rate (*bottom*), which is modulated in phase with the circular eye movement. (Modified from [67])

(eye velocity/target velocity) on the order of 30%, which is usually taken as proof of the importance of the flocculus for pursuit. However, the pursuit deficit is minor compared with the deficit following cerebellectomy. Total cerebellectomy (Fig. 1B, taken from [74], see also [75]) abolishes smooth pursuit completety. Therefore, about 70% of pursuit gain is preserved after flocculectomy! The difference between the effects of cerebellectomy and flocculectomy obviously requires that a major contribution to pursuit is made by cerebellar tissue located outside the flocculus and, perhaps, also the neighboring paraflocculus, since large parts of the paraflocculus were included in the lesions made by Zee et al. [77].

Evidence suggesting that a "supplementary" pursuit center might be located in the posterior vermis was communicated soon after the discovery of "gaze-velocity cells" in the flocculus. In 1979, Kase et al. [31] reported the existence of pursuit-related single units in vermal lobules VI and VII (an example is shown in Fig. 2A), which were thought to encode target velocity rather than gaze velocity. Although the evidence for this interpretation was circumstantial at that time, the concept has been supported by more recent observations on pursuit-related Purkinje cells recorded from the posterior vermis [62, 63]. At first, the discovery of pursuit-related activity received comparatively little attention, being pushed into the background by experiments attributing a major role in saccadic adaptation to the posterior vermis [51]. Such a contribution to saccades fitted nicely into the framework of a division of labor between a pursuit-related cerebellar compartment, the flocculus, and a saccade-related compartment, the posterior vermis.

The interest in a pursuit contribution of the posterior vermis was renewed recently when three laboratories [46, 61, 67] demonstrated pursuit-related activity in the dorsolateral pontine nucleus (DLPN) (see Fig. 2B for an example), which provides a major source of input to the posterior vermis. The demonstration of pursuit-related activity in the DLPN gave rise to the notion that pursuit-related signals originating in the parietooccipital cortex are forwarded to the cerebellum by way of the DLPN, located in the nontegmental base of the pons. Figure 3 presents a diagram of this putative parietooccipito-ponto-cerebellar pathway for pursuit. Although Langer et al. [37] reported on efferent projections connecting this brain-stem nucleus with the flocculus, others [8] have been unable to corroborate this connection. There is compelling evidence, however, from anatomical studies using retrograde tracers injected into various parts of the cerebellum [8, 50, 66], suggesting that the major cerebellar target of fibers originating in the DLPN is the posterior vermis. This view obtains additional support from our own studies using anterograde tracers injected into the DLPN (Thier, unpublished data). Figure 4 shows the distribution of labeled axonal terminals in the cerebellum resulting from an injection of [^{3}H]-proline into the DLPN. Whereas the flocculus and neighboring parts of the cerebellum are virtually devoid of terminals, the posterior vermis and neighboring parts of the hemispheres are heavily labeled. Labeling extends down to lobule IX is a cerebellar center integrating two forms of eye-movement related information using at least two different routes. While saccade-related information is most probably derived from the nucleus reticularis tegmenti pontis (NRTP) [50, 66] (see below for a necessary qualification), we reason that the afferents originating in the DLPN are used to convey pursuit-related signals to the posterior vermis. In view of this double input, it is not surprising to find cells in lobules VI and VII [63] which are modulated by saccades as well as pursuit. The integration of pursuit- and saccade-related information in this part of the posterior vermis is unique among oculomotor centers (of course, excluding the oculomotor nuclei) and the hypothesis is intriguing that the major functional contribution of the "oculomotor vermis" might not be the generation of either saccades or pursuit but their relative coordination. For instance, one may speculate that the posterior vermis may contribute to the control of the size of catch-up saccades needed to accommodate for deficiencies in visual tracking due to a grossly impaired smooth-pursuit component. This view has the advantage that it fits nicely into the framework of adaptive control of saccadic accurracy suggested by earlier work on the posterior vermis [51].

Such a functional role would not necessarily require smooth-pursuit deficiencies to be exhibited after vermal lesions. How can we be sure that the dramatic pursuit deficit following cerebellectomy is a consequence of the absence of a functioning "oculomotor vermis"? While a careful study of pursuit deficiencies following experimental lesions of the posterior vermis has not been carried out, to our knowledge, the admittedly scant clinical data suggest indeed that vermal lesions cause defective smooth-pursuit eye movements in addition to the well-established dysmetria of saccades [53, 55]. Figure 5 shows an example of grossly abnormal smooth pursuit in a patient with an ischemic lesion of large parts of the vermis, while the flocculus and neighboring parts of the cerebellum are spared [52].

Despite the fact that no study of the oculomotor consequences of lesions restricted to the human flocculus has been presented, the prevailing view among clinicians is that floccular lesions in man cause severe pursuit deficits. Although a much more pronounced pursuit deficit in man than in monkeys [77] might reflect a true species difference, it seems more plausible to attribute pursuit disturbances, e.g., in patients with tumors of the cerebellopontine angle, to destruction of tissue relevant for pursuit beyond the flocculus. Obviously, the posterior vermis is too remote to become seriously damaged by a tumor of the cerebellopontine angle. However, there is one structure relevant for pursuit which is close enough to be at risk, namely the base of the pons, including the DLPN and its efferents destined for the posterior vermis. As demonstrated by May et al. [43], chemical inactivation of the dorsolateral pontine nucleus in monkeys causes impaired pursuit in all directions, with some emphasis on ispilaterally directed pursuit (Fig. 6).

A similar pattern of pursuit disturbances, shown in Fig. 7, was recently observed by us [69] in a patient with an ischemic lesion of the dorsal parts of

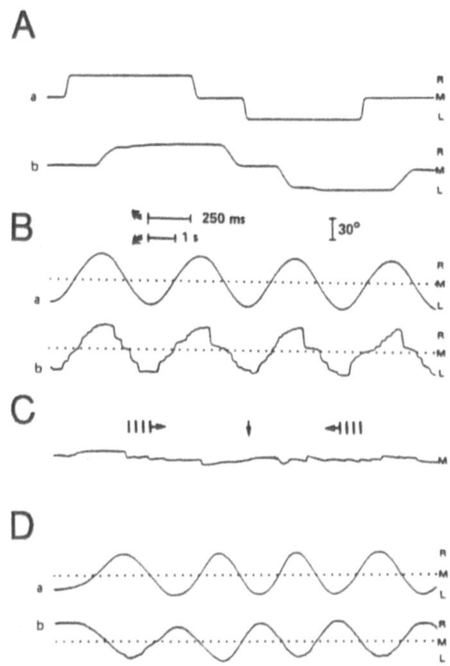

Fig. 5A–D. Smooth-pursuit deficit in man after ischemic lesion of the posterior vermis, sparing the flocculus. **A** Visually guided saccades; **B** smooth pursuit; **C** OKN (rightward stimulation and, after the *vertical arrow*, leftward stimulation); **D** VOR. *a*, target position; *b*, eye position; *R*, right; *M*, straight ahead; *L*, left. [From 52]

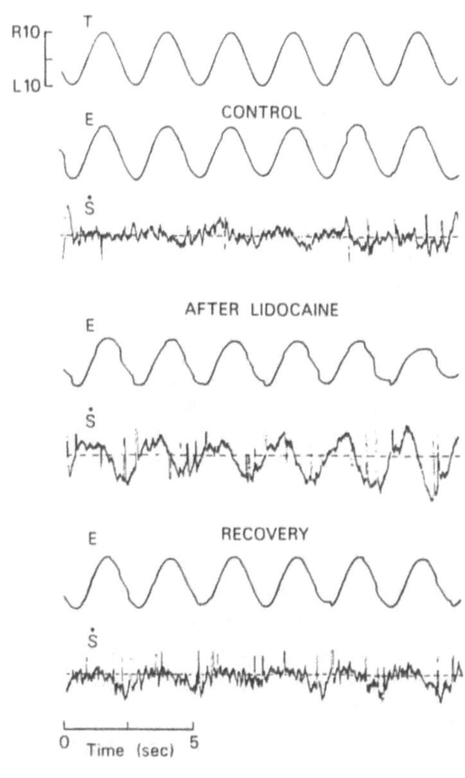

Fig. 6. Deficits in horizontal, sinusoidal smooth-pursuit eye movements in a monkey with a lidocaine lesion of the left dorsolateral pontine nucleus. *T*, target position; *E*, eye position; *s*, retinal slip velocity (saccades removed); *R*, right; *L*, left. [From 43]

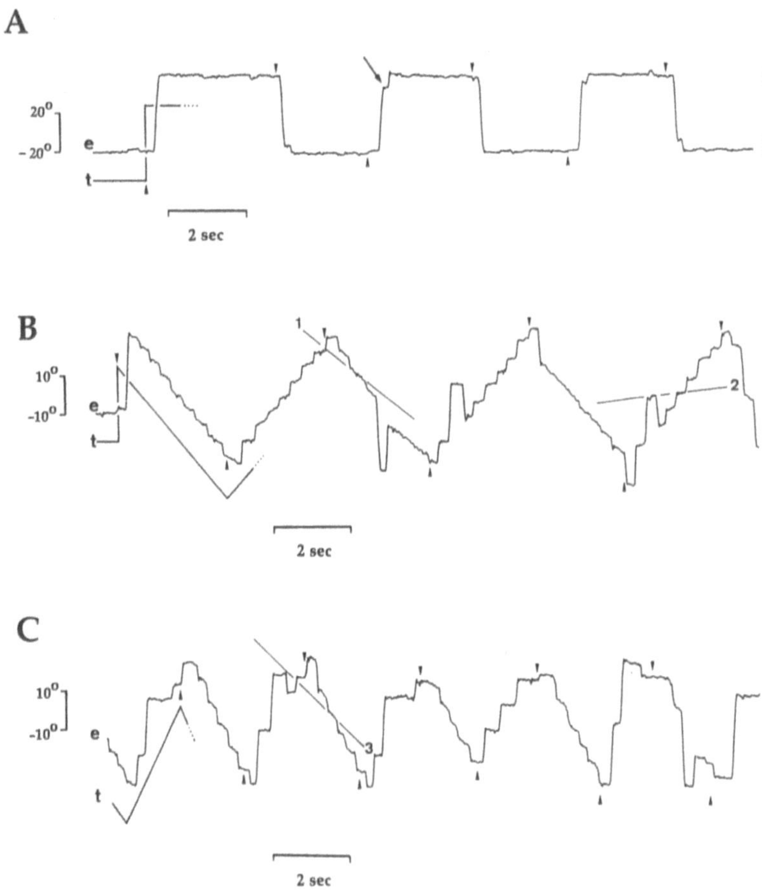

Fig. 7A–C. Voluntary eye movements in a patient with a lesion of the right basal pons. **A** Saccades, elicited by predictable and periodic steps of a small target from left to right and back again. Note occasional hypometric saccades (*arrow* marks an example). **B, C** Visual-tracking eye movements elicited by horizontal movement of a laser target from left to right and back again, according to a triangular waveform. Target velocity was 25°/s in **B** and 40°/s in **C**. *1*, *2*, and *3* are lines indicating mean smooth-pursuit eye velocity. Occurrence of target steps in **A** and turning points of the laser target in **B** and **C** are marked by *small* arrowheads. Note that the samples of the target trajectories (*t*) in **A**, **B**, and **C** are plotted displaced with respect to the eye-position curves (*e*) for the sake of clarity. The electro-oculogram was recorded bitemporally. [From 69]

the basal pons including the DLPN, a deficit which remained unchanged when we reexamined the patient half a year later. In our view, it is a possibility worth considering that at least part of the pursuit deficit observed in patients with tumors of the cerebellopontine angle may be due to damage of neighboring parts of the dorsolateral pons and/or fibers of passage heading for the posterior vermis. Before finishing our discussion of cerebellar contributions to pursuit, we would like to draw the reader's attention to the possibility that the pathway involving the DLPN might not be the only pathway available for pursuit-related signals destined for the cerebellum. Unlike our human case, which failed to show any recovery of the pursuit deficit following a comparatively large lesion of the basal pons, animals with well-restricted experimental lesion of the DLPN-sparing fibers show a remarkable degree of recovery within only a few weeks [43]. The explanation for this discrepancy is that the lesions in monkeys probably spared a parallel pursuit-related pathway affected in the human case. Recent work from the laboratory of Suzuki suggests that this parallel pathway involves the NRTP, located adjacent to the dorsomedial pontine nucleus, hitherto thought to be specialized for saccades [11]. Suzuki and his co-workers [64] were able to demonstrate that the more rostral part of this nucleus contains pursuit-related single-unit activity and, moreover, that microstimulation in this part of the NRTP was able to evoke slow eye movements [76]. Anatomical evidence from our own laboratory (unpublished data) suggests that both the rostral and the caudal parts, i.e., the pursuit- and the saccade-related parts of the NRTP, project indifferently to lobules VIc and VIIA, further emphasizing the notion of an oculomotor vermis which contributes to both forms of visually guided voluntary eye movements.

The integration of saccade-related and pursuit-related information in the posterior vermis is necessary in order to coordinate these two forms of voluntary eye movements. On the other hand, smooth-pursuit eye movements have to be coordinated not only with saccades but also with the nonvolitional, phylogenetically old eye movements to which the flocculus is intimately related [71]. Hence, the presence of pursuit-related activity in the flocculus reflects the need to equip the flocculus with signals required in order to modify "primitive" eye reflexes such as the vestibulo-ocular reflex and the optokinetic reflex, according to the needs of goal-directed eye movements. One particular example of such a coordination between vestibular signals and pursuit-related signals is the suppression of the vestibulo-ocular reflex by pursuit of a target moving with the observer in space. A representation of pursuit-related signals in two segregated cerebellar compartments would therefore reflect phylogenetic needs.

Motor Deficits

Lesions of the Lateral Cerebellum

Acute damage to the lateral cerebellar hemisphere results in hypotonia, hyporeflexia, and asthenia [14, 25]. Unlike patients with acute lesions, those with cerebellar atrophies usually present with normal muscle tone and normal tendon reflexes [16]. Lesions of the lateral cerebellum delay and slow intentional movements, such as grasping or pointing with the affected arm [27]. They are furthermore accompanied by overshoot dysmetria [27]. These abnormalities in timing of movements all contribute to the slowness of rapid alternating movements (dysdiadochokinesia) and to the delayed stop of elbow flexion after release from opposed flexion (impaired check of rebound). Cerebellar tremor may be either static or kinetic. The latter, also called intention tremor or goal-directed tremor, is a sign of disease of the lateral cerebellum or its efferent pathways including the brachium conjunctivum [19]. Tremor emphasizes the proximal parts of the ipsilateral limbs and may be provoked by having the patient perform the finger-nose and the heel-shin tests. The amplitude of the oscillation increases with approach to the target. Static or postural tremor is provoked by asking the patient to hold his arms parallel to the floor or to elevate one leg tonically while standing upright. No information is available as to the site of the lesion responsible for this form of tremor. Myorhythmia [13, 56] is characterized by 2- to 3-Hz oscillations of the palate ("palatal myoclonus"), the tongue, and other muscles innervated by cranial nerves due to a lesion affecting the elements of the triangle of Guillain-Mollaret, consisting of the dentate nucleus, the red nucleus, the central tegmental tract, and the inferior olive. The myorhythmia develops months after the insult and is accompanied by hypertrophy of the inferior olive [35].

Lesions of the Anterior Lobe

Lesions of the anterior lobe, e.g., due to chronic alcoholism [2], cause postural instability characterized by body sway around 3 Hz, predominantly in the anterior-posterior direction [14, 58]. Sway is provoked or enhanced by eye closure [14], proving that visual stabilization of posture is occurring. The postural instability is frequently, but not necessarily, accompanied by an ataxic heel-to-knee test, while ataxia of the upper limb as tested with the finger-to-nose test is absent or mild. Previous work has interpreted the signs of anterior lobe disease to indicate disordered use of spinal feedback information. Without questioning that the anterior lobe is dominated by spinal input, it is worth considering that at least lobules I and II depend on additional input derived from the vestibular complex [4, 5]. It is possible that disruption of vestibular input contributes to the postural instability observed.

Lesions of the Caudal Vermis

Lesions of the caudal vermis including the nodulus, the uvula, and the flocculus (the "vestibulocerebellum", see above), cause postural instability of head and trunk while sitting, standing, and walking. Unlike sway in patients with anterior lobe disease, sway is omnidirectional [14]. Patients with caudal vermis lesions tend to fall even when sitting with their eyes open. Visual stabilization of posture is absent or reduced.

Speech Deficits

Ataxic or cerebellar dysarthria is characterized by slurred pronounciation of consonants and slowed speech tempo [12]. The topography of the underlying lesions is still controversial. Whereas Lechtenberg and Gilman [38] found a significantly higher prevalence of dysarthria in patients with left-sided lesions of the paravermal regions of lobules VI and VII, the observations of Ackermann et al. [1] do not support the notion of a cerebellar speech center being strictly contralateral to the side of the preferred hand. According to the latter authors, purely cortical lesions due to ischemia in the territory of the superior cerebellar artery suffice to cause dysarthria, but exclusively nuclear lesions can probably also cause the articulatory abnormalities.

Nonmotor Functions of the Cerebellum

Traditionally, the approach to the cerebellum has been dominated by the view that its functional contribution can be defined within the framework of motor control. Nevertheless, as reviewed recently by Schmahmann [57] and Leiner et al. [40, 41], even the older clinical literature contains clues to nonmotor functions of the cerebellum. Numerous reports have described disturbances of higher brain functions such as mood disturbances, memory deficits, language deficits not related to deficient articulation, visuospatial difficulties, even psychiatric disease, and many more following cerebellar disease. Based on an examination of the literature, Schmahmann suggested a topographical organization of behavioral and cognitive functions in the cerebellum parallel to the well-established compartimentalization of motor functions. Whereas the more medial parts of the cerebellum are thought to be related to limbic functions such as emotion and sexuality, the more lateral parts are considered to contribute to cognitive functions such as memory, learning, or language. This hypothesis, however, is founded on rather weak grounds, since many of the studies pointing to nonmotor deficits due to cerebellar disease suffer from the fact that pathology was not restricted to the cerebellum. Even more recent studies, adopting a more carefully controlled approach, have not always been able to present noncontroversial interpretations. For example, the view of some authors that the primary site of pathology in children with autism is in vermal lobules VI and VII [6, 10] has been questioned by others [18, 23, 34].

Not only is the topography of nonmotor functions very speculative at present, but also the mechanisms by which the cerebellum might control affective states or cognitive functions remain unclear, and ill-defined concepts [e.g., 22, 41] such as a modulation of the efficacy of cortical communication prevail. There are more specific suggestions, however, such as the idea championed by Keele and Ivry [28, 30, 32] that the cerebellum might contribute to timing processes required by both motor and nonmotor functions. Cognitive deficits such as an impairment of motion perception [29] may therefore be secondary to imprecise timing. Obviously, this is a readaptation of the idea that the cerebellum is an internal clock working in the millisecond range, put forward by Braitenberg and Atwood [7] as early as 1958 on purely anatomical grounds.

In summary, despite the many problems and uncertainties prevailing at the present stage of investigation, the work on nonmotor contributions of the cerebellum clearly has the potential to stimulate future concepts of cerebellar functions and our view of cerebellar compartimentalization.

References

1. Ackermann H, Vogel M, Petersen D, Poremba (1992) Speech deficits in ischemic cerebellar lesions. J Neurol 239:223–227
2. Adams A (1976) Nutritional cerebellar degeneration. In: Vinken PJ, Bruyn GW (eds) Metabolic and deficiency diseases of the nervous system. Elsevier, Amsterdam, pp 271–283 (Handbook of clinical neurology, vol 28)
3. Baloh RW, Jenkins HA, Honrubia V, Yee RD, Lau CGY (1981) Vestibulo-ocular reflex in patients with lesions involving the vestibulocerebellum. Exp Neurol 72: 141–152
4. Barmack NH, Baughman RW, Eckenstein FP (1992) Cholinergic innervation of the cerebellum of rat, rabbit, cat, and monkey as revealed by choline acetyltransferase activity and immunohistochemistry. J Comp Neurol 317:233–249
5. Barmack NH, Baughman RW, Eckenstein FP, Shojaku H (1992) Secondary vestibular cholinergic projection to the cerebellum of rabbit and rat as revealed by choline aceytltransferase immunohistochemistry. Retrograde and orthograde tracers. J Comp Neurol 317:250–270
6. Bauman M, Kemper TL (1985) Histoanatomic observations of the brain in early infantile autism. Neurology (NY) 35:866–874
7. Braitenberg V, Atwood RP (1958) Morphological observations on the cerebellar cortex. J Comp Neurol 109:1–33
8. Brodal P (1982) Further observations on the cerebellar projections from the pontine nuclei and the nucleus reticularis tegmenti pontis in the rhesus monkey. J Comp Neurol 204:44–55
9. Cannon SC, Robinson (1987) Loss of the neural integrator of the oculomotor system from brain-stem lesions in monkey. J Neurophysiol 57:1383–1409
10. Courchesne E (1991) Neuroanatomic imaging in autism. Pediatrics 87:781–790
11. Crandall WF, Keller EL (1985) Visual and oculomotor signals in NRTP in alert monkey. J Neurophysiol 54:1326–1345
12. Darley FL, Aronson AE, Brown JR (1975) Motor speech disorders. Saunders, Philadelphia
13. Deuschl G, Mischke G, Schenck E, Schulte-Mönting, Lücking CH (1990) Symptomatic and essential rhythmic palatal myoclonus. Brain 113:1645–1672
14. Dichgans J, Diener HC (1984) Clinical evidence for functional compartmentalization of the cerebellum. In: Bloedel JR, Dichgans J, Precht W (eds) Cerebellar functions. Springer, Berlin Heidelberg New York, pp 126–147
15. Dichgans J, von Reutern GM, Römmelt U (1978) Impaired suppression of vestibular nystagmus by fixation in cerebellar and noncerebellar patients. Arch Psychiatr Nervenkr 226:183–199
16. Diener H-C, Dichgans J (1992) Pathophysiology of cerebellar ataxia in humans. Mov Disord 7:95–109

17. Edwards SB, Hendrickson A (1981) The autoradiographic tracing of axonal connections in the central nervous system. In: Heimer L, Robards MJ (eds) Neuroanatomical tract-tracing methods. Plenum, New York, pp 71–206
18. Ekman et al. (1991) Low field magnetic resonance imaging of the central nervous system in 15 children with autistic disorder. Acta Paediatr Scand 80:243–247
19. Gilman S, Bloedel J, Lechtenberg R (1981) Disorders of cerebellum. Davis, Philadelphia
20. Hain TC, Zee DS, Maria BL (1988) Tilt suppression of vestibulo-ocular reflex in patients with cerebellar lesions. Acta Otolaryngol (Stockh) 105:13–20
21. Halmagy GM, Rudge P, Gresty MA, Sanders MD (1983) Downbeating nystagmus. A review of 62 cases. Arch Neurol 40:777–784
22. Heath RJ (1977) Modulation of emotion with a brain pacemaker. J Nerv Ment Dis 165:300–317
23. Heh CW, Smith R, Wu J, Hazlett E, Rusell A, Asarnow R, Tanguay P, Buchsbaum MS (1989) Positron emission tomography of the cerebellum in autism. Am J Psychiatry 146:242–245
24. Heinen SJ, Keller EL (1990) Cerebellar uvula correlates of moving visual stimuli. Soc Neurosci Abstr 16:1083
25. Holmes G (1917) The symptoms of acute cerebellar injuries due to gunshot injuries. Brain 40:461–535
26. Holmes G (1930) The cerebellum of man. Brain 62:1–30
27. Hore J, Wild B, Diener HC (1991) Cerebellar dysmetria at the elbow, wrist and fingers. J Neurcophysiol 65:563–571
28. Ivry RB, Baldo JV (1992) Is the cerebellum involved in learning and cognition? Curr Opin Neurobiol 2:212–216
29. Ivry RB, Diener HC (1991) Impaired velocity perception in patients with lesions of the cerebellum. J Cogn Neurosci 3:355–366
30. Ivry RB, Keele SW (1989) Timing functions of the cerebellum. J Cogn Neurosci 1:134–150
31. Kase M, Noda H, Suzuki DA, Miller DC (1979) Target velocity signals of visual tracking in vermal Purkinje cells of the monkey. Science 203:717–720
32. Keele SW, Ivry R (1990) Does the cerebellum provide a common computation for diverse tasks? Ann NY Acad Sci 608:179–211
33. Keller EL (1989) The cerebellum. In: Wurtz, Goldberg (eds) The neurobiology of saccadic eye movements. Elsevier, Amsterdam
34. Kleimann MD, Neff S, Rosman NP (1992) The brain in infantile autism: are posterior fossa structures abnormal? Neurology 42:753–760
35. Koeppen AH, Barron KD, Dentinger MP (1980) Olivary hypertrophy in man. In: Courville J, de Montigny C, Lamarre Y (eds) The inferior olivary nucleus. Anatomy and physiology. Raven, New York, pp 309–314
36. Langer T, Fuchs AF, Scudder CA, Chubb MC (1985) Afferents to the flocculus of the cerebellum in the rhesus macaque as revealed by retrograde transport of horseadish peroxidase. J Comp Neurol 235:1–25
37. Langer T, Fuchs AF, Chubb MC, Scudder CA, Lisberger SG (1985) Floccular efferents in the rhesus macaque as revealed by autoradiography and horseradish peroxidase J Comp Neurol 235:26–37
38. Lechtenberg R, Gilman S (1978) Speech disorders in cerebellar disease. Ann Neurol 3:285–290
39. Leigh RJ, Robinson DA, Zee DS (1981) A hypothetical explanation for periodic alternating nystagmus: instability in the optokinetic-vestibular system. Ann NY Acad Sci 374:619–635
40. Leiner HC, Leiner AL, Dow RS (1986) Does the cerebellum contribute to mental skills? Behav Neurosci 100:443–454
41. Leiner HC, Leiner AL, Dow RS (1991) The human cerebro-cerebellar system: its computing, cognitive, and language skills. Behav Brain Res 44:113–128
42. Lisberger SG, Fuchs AF (1978) Role of primate flocculus during rapid behavioral modification of vestibulo-ocular reflex. I. Purkinje cell activity during visually guided

horizontal smooth-pursuit eye movements and passive head rotation. J Neurophysiol 41:733–763
43. May JG, Keller EL, Suzuki DA (1988) Smooth pursuit eye movement deficits with chemical lesions in the dorsolateral pontine nucleus of the monkey. J Neurophysiol 59:952–977
44. Miles FA, Fuller JH (1975) Visual tracking and the primate flocculus. Science 189: 1000–1002
45. Miles FA, Fuller JH, Braitman DJ, Dow BM (1980) Long-term adaptive changes in primate vestibulo-ocular reflex. III. Electrophysiological observations in flocculus of normal monkeys. J Neurophysiol 43:1437–1476
46. Mustari MJ, Fuchs AF, Wallman J (1988) Response properties of dorsolateral pontine units during smooth pursuit in the rhesus macaque. J Neurophysiol 60:664–686
47. Newsome WT, Wurtz RH (1988) Probing visual cortical function with discrete chemical lesions. TINS 11:394–399
48. Noda H, Fujikado T (1987) Topography of the oculomotor area of the cerebellar vermis in macaques as determined by microstimulation. J Neurophysiol 58:359–378
49. Noda H, Fujikado T (1987) Involvement of Purkinje cells in evoking saccadic eye movments by microstimulation of the posterior cerebellar vermis of monkeys. J Neurophysiol 57:1247–1261
50. Noda H, Sugita S, Ikeda Y (1990) Afferent and efferent connections of the oculomotor region of the fastigial nucleus in the macaque monkey. J Comp Neurol 302:330–348
51. Optican LM, Robinson DA (1980) Cerebellar-dependent adaptive control of the primate saccadic system. J Neurophysiol 44:1058–1076
52. Pierrot-Deseilligny C, Amarenco P, Roullet E, Marteau R (1990) Vermal infarct with pursuit eye movement disorders. J Neurol Neurosurg Psychiatry 53:519–521
53. Ritchie L (1976) Effects of cerebellar lesions on saccadic eye movements. J Neurophysiol 39:1246–1256
54. Robinson DA (1981) Control of eye movements. In: Brookhart JM, Mountcastle VB (eds) The nervous system, vol 2. American Physiological Society, Bethesda, pp 1275–1320 (Handbook of physiology, sect 1)
55. Sato H, Noda H (1992) Saccadic dysmetria induced by transient functional decoration of the cerebellar vermis. Exp Brain Res 88:455–458
56. Schenck E (1965) Die Hirnnervenmyorhythmie, ihre Pathogenese und ihre Stellung im myoklonischen Syndrom. Springer, Berlin Heidelberg New York (Monographien aus dem Gesamtgebiete der Neurologie und Psychiatrie, Heft 109)
57. Schmahmann JD (1991) An emerging concept. The cerebellar contribution to higher function. Arch Neurol 48:1178–1187
58. Silfverskiöld BP (1969) Romberg's test in the cerebellar syndrome occurring in chronic alcoholism. Acta Neurol Scand 45:292–302
59. Stone LS, Lisberger SG (1990) Visual responses of Purkinje cells in the cerebellar flocculus during smooth-pursuit eye movements in monkeys. I. Simple spikes. J Neurophysiol 63:1241–1261
60. Stone LS, Lisberger SG (1990) Visual responses of Purkinje cells in the cerebellar flocculus during smooth-pursuit eye movements in monkeys. II. Complex spikes. J Neurophysiol 63:1262–1275
61. Suzuki DA, Keller EL (1984) Visual signals in the dorsolateral pontine nucleus of the alert monkey: their relationship to smooth-pursuit eye movements. Exp Brain Res 53:473–478
62. Suzuki DA, Keller EL (1988) The role of the posterior vermis of monkey cerebellum in smooth-pursuit eye movement control. I. Eye and head movement-related activity. J Neurophysiol 59:1–17
63. Suzuki DA, Keller EL (1988) The role of the posterior vermis of monkey cerebellum in smooth-pursuit eye movement control. II. Target velocity-related Purkinje cell activity. J Neurophysiol 59:19–40

64. Suzuki DA, Yamada T, Betelak KF, Yee RD (1991) Smooth-pursuit eye movement-related activity in monkey nucleus reticularis tegmenti pontis. Soc Neurosci Abstr 17:459
65. Takemori S, Cohen B (1974) Loss of suppression of vestibular nystagmus after flocculus lesion. Brain Res 72:213–224
66. Thielert C-D, Thier P (1992) A comparison of the patterns of projections from the pontine nuclei and the nucleus reticularis tegmenti pontis to the "oculomotor" vermis in the monkey: a study using retrograde tracers. Rhythmogenesis in neurons and networks. Proceedings of the 20th Göttingen Neurobiology Conference. Thieme, Stuttgart, p 410
67. Thier P, Koehler W, Buettner UW (1988) Neuronal activity in the dorsolateral pontine nucleus of the alert monkey modulated by visual stimuli and eye movements. Exp Brain Res 70:496–512
68. Thier P, Koehler W, Buettner UW, Dichgans J (1990) Does the system for smooth-pursuit eye movements rely on a neuronal representation of target motion in space? In: Deecke L, Eccles JC, Mountcastle VB (eds) From neuron to action. An appraisal of fundamental and clinical research. Springer, Berlin Heidelberg New York, pp 133–142
69. Thier P, Bachor A, Faiss J, Dichgans J. Koenig E (1991) Selective impairment of smooth-pursuit eye movements due to an ischemic lesion of the basal pons. Ann Neurol 29:443–448
70. Von Reutern GM, Dichgans J (1977) Augenbewegungstörungen als cerebelläre Symptome bei Kleinhirnbrückenwinkeltumoren. Arch Psychiatr Nervenkr 223:117–130
71. Waespe W, Henn V (1984) The primate flocculus in visual-vestibular interactions: conceptual, neurophysiological, and anatomical problems. In: Bloedel JR, Dichgans J, Precht W (eds) Cerebellar functions. Springer, Berlin Heidelberg New York
72. Waespe W, Cohen B, Raphan T (1985) Dynamic modification of the vestibulo-ocular reflex by the nodulus and uvula. Science 228:199–202
73. Walberg F, Dietrichs E (1988) The interconnection between the vestibular nuclei and the nodulus: a study of reciprocity. Brain Res 449:47–53
74. Westheimer G, Blair SM (1973) Oculomotor defects in cerebellectomized monkeys. Invest Ophthalmol 12:618–621
75. Westheimer G, Blair SM (1974) Functional organization of primate oculomotor system revealed by cerebellectomy. Exp Brain Res 21 21:463–472
76. Yamada T, Suzuki DA, Betelak KF, Yee RD (1991) Initial eye position dependence of eye movements evoked by microstimulation in the monkey nucleus reticularis tegmenti pontis. Soc Neurosci Abstr 17:459
77. Zee DS, Yamazaki A, Butler PH, Gücer G (1981) Effects of ablation of flocculus and paraflocculus on eye movements in primate. J Neurophysiol 46:878–899

Is Saccadic Lateropulsion in Wallenberg's Syndrome Caused by a Cerebellar or a Brain-Stem Lesion?

A. Straube[1], C. Helmchen[1], V.R. Robinson[2], A. Fuchs[2], and U. Büttner[1]

[1] Neurologische Klinik, Klinikum Großhadern, Ludwig Maximilians-Universität, Marchioninistraße 15, 81377 München, Germany
[2] Institute of Physiology and Biophysics, Regional Primate Center, University of Washington, Seattle, WA, USA

Introduction

One of the common features of Wallenberg's lateral medullary syndrome is the tendency to fall towards the side of the lesion, called lateropulsion [4]. Less well known, however, is ocular lateropulsion [9], i.e., a tonic drift of the eyes to the ipsilateral side, compatible with the direction of the body lateropulsion. Also, there is a directional preponderance of saccades to the side of the lesion [1, 10] affecting quick phases of vestibular and optokinetic nystagmus [9]. Thus, saccades to the contralateral side of the lesion, i.e., against the ocular bias, are hypometric, whereas ipsilateral saccades are hypermetric. Other oculomotor signs include cogwheel smooth-pursuit eye movements directed against the lateropulsive bias, i.e., to the contralateral side [1], and the ocular tilt reaction with skew deviation [6].

Direction-specific saccadic dysmetria, however, is quite a common sign in cerebellar disease. Since Wallenberg's syndrome is usually due to occlusion of the vertebral artery or infarction in the territory of the posterior inferior cerebellar artery [7], both the brainstem and the cerebellum can be involved, and it has remained unclear which part is responsible for the saccadic lateropulsion. Disruption of cerebellar pathways between the paravermian region and the paramedian pontine reticular formation (PPRF) has been suggested to explain the ocular position preponderance [1]. Others have postulated that saccadic lateropulsion is due to a destruction of afferent fibers to the cerebellum within the brain stem [12].

Therefore, we have started a prospective study on saccadic lateropulsion in patients with unilateral cerebellar disease and lateral medullary infarction, based on clinical examination, electro-oculographic eye movement recordings, and localization of the lesion by CT scans and MRI. Saccadic eye movements in these patients are compared with visually guided saccades in the monkey after selective reversible lesions in the deep cerebellar nuclei, causing saccadic dysmetria. Preliminary data obtained following microinjections of the GABAergic substance muscimol in the fastigial nucleus of the monkey show that this substance can apparently cause severe saccadic and body lateropulsion to the affected side (unpublished data). The topodiag-

L.R. Caplan and H.C. Hopf (Eds.)
Brain-Stem Localization and Function

nostic localization of the lesion (brain stem versus cerebellum) would help to elucidate the underlying pathomechanisms of saccadic lateropulsion.

Methods

Patients of all ages are included into the study if they show at least three typical clinical signs of Wallenberg's lateral medullary syndrome (Horner's syndrome, ipsilateral ataxia, contralateral loss of pain and temperature sensation, ipsilateral facial loss of sensation, spontaneous nystagmus, dysarthria and/or dysphagia). Moreover, CT scans and/or MRI are used to identify cerebellar infarction and/or a unilateral lesion in the lateral medulla oblongata.

Horizontal and vertical eye movements in patients are recorded by electro-oculography or infrared oculography as soon as possible (hours to days) after the onset of Wallenberg's syndrome and in subsequent examinations. Targets are presented in the light as red LED spots on a convex screen. Horizontal amplitudes include 10°, 20°, 40°, 60°, and 80° target steps. The amplitudes of at least 20 saccades (10 saccades in each direction, left and right) are averaged to determine the gain (amplitude eye movement/ amplitude target step). Also, subsequent corrective saccades are quantitatively analyzed. Normal reference data are used from 20 healthy persons, recorded in the same laboratory [2]. The gain of the control group is 0.937 (SD 0.01 for 10° and 20° saccades).

Horizontal and vertical eye position in trained monkeys (*Macaca fascicularis*) are recorded by the scleral search coil method [8]. The fastigial nucleus was electrophysiologically localized. Microinjections of 0.5–1.0 μg muscimol, a GABA-A-agonist, in 0.5–1.0 μl NaCl (0.9%) are given unilaterally into the caudal fastigial nucleus, with subsequent recording of spontaneous and visually guided saccades for at least 1 h.

Results

Preliminary data show saccadic dysmetria in patients with Wallenberg's syndrome and in monkeys after unilateral microinjection of the GABAergic agent muscimol into the fastigial nucleus. In both man and monkey, horizontal saccades to the contralateral side of the lesion were hypometric, while ipsilateral saccades were hypermetric. Figure 1 shows the recording of a 29-year-old patient with Wallenberg's syndrome on the right, with hypermetric saccades (gain 1.25) to the right and hypometric saccades (gain 0.29) to the left. The gain in the eight patients ranged from 0.89 to 1.25 for ipsilateral saccades (mean gain: 1.00) and from 0.29 to 0.69 for contralateral saccades (mean gain: 0.58). Likewise, saccadic eye movements of the monkey towards the side of the microinjection into the fastigial nucleus were hypermetric and contralateral, saccades were hypometric (Fig. 2).

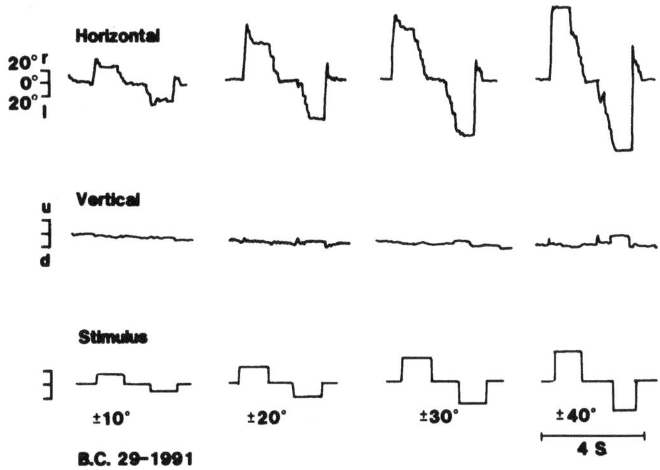

Fig. 1. Recording of a 29-year-old patient with Wallenberg's syndrome on the right: saccades to the right (ipsilateral) are hypermetric, saccades to the left (contralateral) hypometric. *Upper line* shows horizontal saccades with increasing amplitude (10°, 20°, 30°, 40°). *Bottom line* indicates the stimulus; the *middle line* represents the vertical eye position. *r*, right; *l*, left; *u*, up; *d*, down

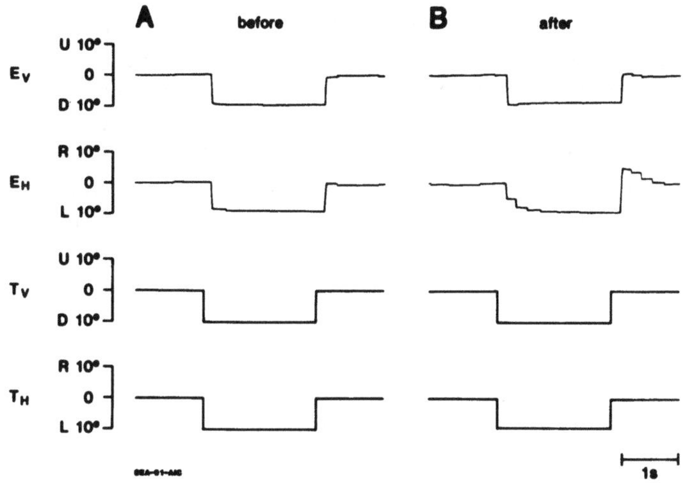

Fig. 2 A,B. Saccades in the *monkey* before (**A**) and after (**B**) microinjection of muscimol (1 µg/µl) in the right fastigial nucleus. *Upper two lines* in **A** show fixating vertical (E_V) and horizontal (E_H) eye movements following the stimulus (T_V and T_H, respectively) given below with + /− 10°. After muscimol microinjection (**B**), saccades are hypermetric to the right (ipsilateral) and hypometric to the left (contralateral)

Discussion

Oculomotor signs are obligatory in Wallenberg's syndrome [11], and their hallmarks are saccadic lateropulsion and a smooth-pursuit deficit towards the contralateral side. None of our patients had a lesion in the deep cerebellar nuclei. However, qualitatively and quantitatively similar disturbances of eye movements can be found after unilateral muscimol microinjections into the medial deep cerebellar nuclei of the macaque monkey. Moreover, the monkeys showed a tendency to fall towards the side of the lesion without accompanying signs of limb ataxia comparable to the body lateropulsion of patients with Wallenberg's syndrome.

What is the common mechanism for the oculomotor signs in Wallenberg's syndrome and after unilateral lesions of the fastigial nucleus? Kommerell and Hoyt [9] and, more recently, Waespe and Wichmann [12] put forward the hypothesis that a lesion of cerebellar afferents within the brain stem is the cause of the oculomotor signs in Wallenberg's syndrome. The vascular lesion in the lateral medullary syndrome regularly affects the olivocerebellar fibers (climbing fibers) which cross at this level from the contralateral inferior olive and travel through the inferior cerebellar peduncle to the cerebellar cortex. Physiologically, activation of these climbing fibers causes a complex spike activity in Purkinje cells with long-lasting hyperpolarization, which leads to a decrease of net simple spike activity by 10%. However, after the destruction of the climbing fibers in the lateral medullary brain stem simple spike activity of Purkinje cells is increased by more than 100% due to functional disinhibition. As a consequence, the physiologically GABA-mediated inhibition of the deep cerebellar nuclei is released, which can be compared with the effect of a microinjection of the GABA agonist muscimol into the deep cerebellar nuclei [3, 5]. In summary, saccadic lateropulsion, the smooth-pursuit deficit, and possibly the lateropulsion of the body can result from a lesion of the afferent climbing fibers to the cerebellar cortex.

References

1. Baloh RW, Yee RD, Honrubia V (1981) Eye movements in patients with Wallenberg's syndrome. Ann NY Acad Sci 374:600–613
2. Bötzel K, Rottach K, Büttner U (1993) Normal and pathological saccadic dysmetria. Brain 116:337–353
3. Benedetti F, Montarolo PG, Strata P, Tempia F (1983) Inferior olive inactivation decreases the excitability of the intracerebellar and lateral vestibular nuclei in the rat. J Physiol (Lond) 340:195–208
4. Bjerver K, Silfverskiöld BP (1968) Lateropulsion and imbalance in Wallenberg's syndrome. Acta Neurol Scand 44:91–100
5. Demer JL, Echelman DA, Robinson DA (1985) Effects of electrical stimulation and reversible lesions of the olivocerebellar pathway on Purkinje cell activity in the flocculus of the cat. Brain Res 346:22–31

6. Dieterich M, Brandt T (1992) Wallenberg's syndrome: lateropulsion, cyclorotation, and subjective visual vertical in thirty-six patients. Ann Neurol 31:399–408
7. Fisher CM, Karnes WE, Kubik CS (1961) Lateral medullary infarction – the pattern of vascular occlusion. J Neuropathol Exp Neurol 20:323–379
8. Judge SJ, Richmond BJ, Chu FC (1980) Implantation of magnetic search coils for measurements of eye position: an improved method. Vision 20:535–538
9. Kommerell G, Hoyt WF (1973) Lateropulsion of saccadic eye movements. Arch Neurol 28:313–318
10. Meyer KT, Baloh RW, Krohel GB, Hepler RS (1980) Ocular lateropulsion: a sign of lateral medullary disease. Arch Ophthalmol 98:1614–1616
11. Norrving B, Cronqvist S (1991) Lateral medullary infarction: prognosis in an unselected series. Neurology 41:244–248
12. Waespe W, Wichmann W (1990) Oculomotor disturbances during visual-vestibular interaction in Wallenberg's lateral medullary syndrome. Brain 113:821–846

Paroxysmal Ataxia and Dysarthria with a Single Lesion in the Cerebellar Peduncle

T. Kammer, D. Linden, R.R. Diehl, and M. Hennerici

Abteilung für Neurologie, Klinikum Mannheim, Universität Heidelberg, Theodor-Kutzer-Ufer, 68167 Mannheim, Germany

Introduction

Paroxysmal neurological disturbances lasting only seconds are well known in multiple sclerosis (MS) [4, 6]. The main types which have been reported are trigeminal neuralgia, pain attacks of the limbs, tonic seizures, paroxysmal paresthesia (e.g., sign of Lhermitte), ataxia, and dysarthria. The incidence of most paroxysmal symptoms in MS ranges from 3%–5% [2] up to an incidence of 17% for tonic seizures in MS observed in Japan [9].

Paroxysmal hemiataxia and dysarthria are reported to be characteristic of MS or even pathognomonic [11]; they can be the first manifestations of the disease [10]. However, the localization of the lesion has not been established in a single case, although most investigators hypothesize a brain-stem or cerebellum localization [5]. With the availability of magnetic resonance imaging (MRI), a detailed localization is possible: We report the first case of paroxysmal ataxia and dysarthria with clinically defined MS and a singular lesion on MRI.

Case Report

A 56-year-old housewife presented with a 6-week history of recurrent feelings of lightness of her left arm and leg with simultaneous dizziness and slurred, hesitant speech. Movement control of the left side was abnormal during the attacks. Each episode occurred without warning but they were sometimes triggered by rapid head movements. The duration was always less than 10 s and the frequency was up to 20 per day.

There was a history of transverse myelitis with paraparesis at the age of 46 years. MRI of the spinal cord showed a T_2-weighted intensity from C-3 to C-6. Pathological findings did not occur in either cerebrospinal fluid (CSF) or evoked potentials. After several courses of treatment with ACTH and dexamethasone, a complete remission occurred 2 years after onset.

Neurological examinations were completely normal during the intervals between the attacks. CT scans with and without contrast medium were

L.R. Caplan and H.C. Hopf (Eds.)
Brain-Stem Localization and Function

References

1. Berger JR, Sheremata WA, Melamed E (1984) Paroxysmal dystonia as the initial manifestation of multiple sclerosis. Arch Neurol 41:747–750
2. Espir MLE, Millac P (1970) Treatment of paroxysmal disorders in multiple sclerosis with carbamazepine (Tegretol). J Neurol Neurosurg Psychiatry 33:528–531
3. Gorard DA, Gibberd FB (1989) Paroxysmal dysarthria and ataxia: associated MRI abnormality. J Neurol Neurosurg Psychiatry 52:1444–1445
4. Hopf HC, Stamatovic AM, Wahren W (1970) Die cerebralen Anfälle bei der multiplen Sklerose. Z Neurol 198:256–279
5. Matthews WB (1975) Paroxysmal symptoms in multiple sclerosis. J Neurol Neurosurg Psychiatry 38:617–623
6. Matthews WB (ed) (1991) McAlpine's multiple sclerosis, 2nd edn. Churchill Livingstone, Edinburgh; p 64
7. Osterman PO, Westerberg CE (1975) Paroxysmal attacks in multiple sclerosis. Brain 98:189–202
8. Poser CM, Paty DW, Scheinberg L, McDonald WI, Davis FA, Ebers GC, Johnson KP, Sibley WA, Silberberg DH, Tourtellotte WW (1983) New diagnostic criteria for multiple sclerosis: guidelines for research protocols. Ann Neurol 13:277–231
9. Shibasaki H, Kuroiwa Y (1974) Painful tonic seizures in multiple sclerosis. Arch Neurol 30:47–51
10. Twomey JA, Espir MLE (1980) Paroxysmal symptoms as the first manifestation of multiple sclerosis. J Neurol Neurosurg Psychiatry 43:296–304
11. Wolf P, Assmus H (1974) Paroxysmale Dysarthrie und Ataxie. Ein pathognomonisches Anfallssyndrom bei multipler Sklerose. J Neurol 208:27–38

Preliminary Classification of Vestibular Brain-Stem Disorders

T. Brandt and M. Dieterich

Neurologische Klinik, Klinikum Großhadern, Ludwig Maximilians-Universität, Marchioninistraße 15, 81377 München, Germany

It is well recognized that some of the supranuclear ocular motor disorders are secondary to distinct and separated lesions of central vestibular pathways from the vestibular nuclei to ocular motor nuclei. A classification of central vestibular disorders of the brain stem and vestibulocerebellum does not exist. Therefore, an attempt will be made to attribute downbeat/upbeat nystagmus, pseudo-vestibular neuritis, and ocular tilt reaction (OTR) to lesional tone imbalances in one of the three major planes of action of the vestibuloocular reflex (VOR).

We regard this as a first hypothetical and speculative attempt to classify vestibular brain-stem disorders [3]. It is based on clinical experience and on animal studies showing that the same syndrome, such as downbeat or upbeat nystagmus or OTR, can be elicited by circumscribed lesions of vestibular pathways at different brain-stem levels from the medulla to the mesencephalon. A prerequisite is that each pathway that mediates the VOR in one of the three major planes runs a separate course in order to become individually lesioned. Such a classification is helpful for the clinician in terms of pathophysiological thinking and topological diagnosis. A rigid three-plane scheme is, however, insufficient to cover our current knowledge of a complex vestibular neuronal network.

The Three-Dimensional VOR

The VOR normally serves to hold the direction of gaze in space constant during head movements, by forcing the eyes to move in their orbits in the direction opposite to that of head motion with a velocity and amplitude which "compensates" for the head motion. The VOR has three major planes of action:

1. Horizontal head rotation about a vertical z-axis = yaw
2. Head extension and flexion about a horizontal y-axis = pitch
3. Lateral head tilt about a horizontal x-axis = roll

L.R. Caplan and H.C. Hopf (Eds.)
Brain-Stem Localization and Function

These planes represent three-dimensional space in the vestibular and oculomotor system, with different demands upon the VOR for perception of a stable world and postural balance during head rotations in yaw, pitch, or roll.

The neuronal network mediating VOR of horizontal and vertical semicircular canals is based upon sensory convergence within a three-neuron reflex arc. It links a set of extraocular muscles with their primary action, roughly aligned to the particular spatial plane of either the horizontal, anterior, or posterior canals. Sensory-motor transformation from canal planes to the planes of eye movements has been demonstrated at the level of second-

Table 1. Central vestibular vertigo syndromes

Site	Syndrome	Mechanism/etiology
Vestibular cortex	Vestibular epilepsy	Simple partial seizures
Thalamus (interstitial nucleus of Cajal)	Ocular tilt reaction	Paramedian thalamic infarction, hemorrhage
Vestibulocerebellum	Thalamic astasia	Posturolateral thalamic lesions
Flocculus	Downbeat nystagmus/vertigo	Disinhibition of vertical VOR in pitch
Nodulus	Positional downbeat nystagmus	Inappropriate otolith–canal interaction?
Vestibular nuclei; vestibulocerebellar loop?	Central positional vertigo and/or nystagmus	Inappropriate otolith–canal interaction?
Mesodiencephalic brain-stem	Ocular tilt reaction	Tone imbalance of VOR in roll
	Upbeat nystagmus/vertigo	Tone imbalance of vertical VOR in pitch
Pontomedullary brain-stem	Paroxysmal dysarthria/ataxia in MS	Transversely spreading ephaptic axonal activation
	Pseudo-"vestibular neuritis"	Vestibular nerve root plaque in MS or lacunar infarction
	Ocular tilt reaction	Graviceptive pathway lesion with tone imbalance of VOR in roll
	Downbeat nystagmus/vertigo	Tone imbalance of VOR in pitch
Medullary brain-stem	Upbeat nystagmus/vertigo	Tone imbalance of VOR in pitch
	Paroxysmal vertigo evoked by lateral gaze or head position	Vestibular nuclei lesion?
Brain-stem and cerebellum	Pseudo-"vestibular neuritis"	Ischemia of superior labyrinth, vestibular nerve, vestibular nuclei (AICA or PICA infarction)
	Lateropulsion in Wallenberg's syndrome	Ischemic lesion-induced deviation of subjective vertical
	Familial periodic vertigo	Hereditary, metabolic?
	Encephalitis with predominant vertigo	Virus infection
	Epidemic vertigo	Virus infection

order neurons within the vestibular nuclei [17]. The three-neuron arc of the VOR is essential for the short latency response; the overall capacity of the VOR most likely depends upon a distributed, parallel network involving vestibular nuclei neurons, as described by Anastasio and Robinson [1].

When the head is turned, all semicircular canals are stimulated. However, VOR in yaw is mediated predominantly by the horizontal canals and VOR in pitch and roll by the vertical anterior and posterior canals. Spatial organization of right and left labyrinths in temporal bones is such that horizontal and vertical canals can be paired with respect to their optimal plane of function although the canals are not completely coplanar.

The common view that the VOR merely serves stabilization of gaze should be corrected. Its neuronal pathways also include ascending input to the thalamocortical projections (Fig. 8) of perception as well as descending input to vestibulospinal projections for adjustments of head and body posture (vestibulospinal reflexes). Thus, the vestibular nuclei output of the VOR branches, to reach the vestibular cortex areas for perception and the spinal cord for posture. This means that both physiologic stimulation or pathologic dysfunction of the semicircular canal input not only provoke nystagmus but inevitably cause a direction-specific concurrent rotary vertigo and postural imbalance. The VOR is not an independent oculomotor reflex, but an integral part of the sensorimotor vestibular response [22]. With respect to the neuronal linkage, VOR is inconceivable without concurrent postural and perceptual correlates, both under experimental conditions and in disease.

Brain-stem Disorders of the VOR

Central vestibular vertigo syndromes are caused in the majority of cases by dysfunction induced by a lesion, but a small proportion result from pathologic excitation of various structures, ranging from the vestibular nuclei

Table 2. A reasonable classification of some central vestibular syndromes of the brain-stem tegmentum

Disorders of the VOR in horizontal (yaw) plane	– Horizontal nystagmus • pseudo-"vestibular neuritis" (partial AICA/PICA infarctions; MS plaques)
Disorders of the VOR in sagittal (pitch) plane	– Vertical nystagmus • downbeat nystagmus • upbeat nystagmus
Disorders of the VOR in frontal (roll) plane	– ocular torsion • ocular tilt reaction (lateropulsion)

AICA, anterior inferior cerebellar artery; PICA, posterior inferior cerebellar artery; MS, multiple sclerosis.

to the vestibular cortex (Table 1). Most patients have intra-axial, paramedian infratentorial lesions, especially of the pontomedullary tegmentum involving vestibular nuclei connections to vestibulocerebellar structures.

Evidence is presented for a reasonable classification of some central vestibular syndromes of the brain-stem tegmentum in Table 2.

Pseudo-Vestibular Neuritis

Most often, vestibular neuritis is caused by viral infection of the vestibular nerve and is most probably a partial rather than a complete vestibular paresis with predominant involvement of the horizontal semicircular canal [7]. The chief symptom is the acute onset of prolonged severe rotational vertigo, associated with spontaneous contraversive nystagmus, postural imbalance, and nausea. Caloric testing invariably shows ipsilateral hypo- or nonresponsiveness (horizontal semicircular canal paresis).

Unilateral loss of vestibular function can be produced by conditions other than the common vestibular neuritis. Small infarctions in the territory of either the anterior inferior cerebellar artery (AICA) or the posterior inferior cerebellar artery (PICA) have been repeatedly reported to mimic unilateral peripheral labyrinthine disorder, mostly with a similarly benign course [13, 25]. Hopf [19] has described patients with additional impairment of the masseter reflex and argued that a small region ventrolateral to the floor

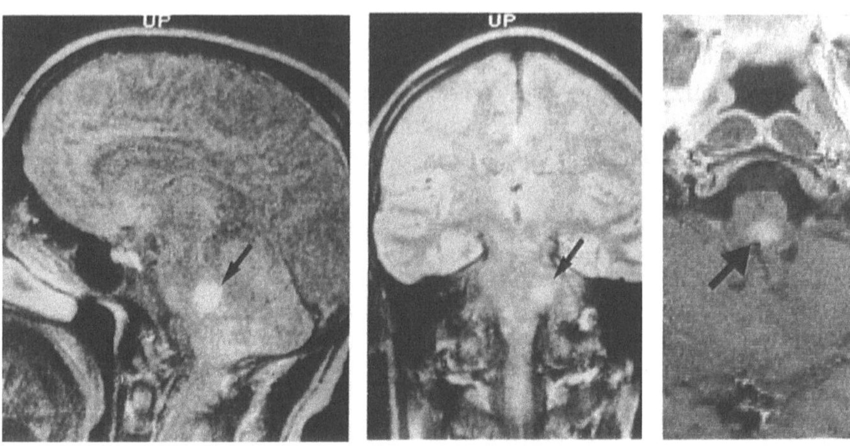

Fig. 1. Central (pseudo) vestibular neuritis in two patients with multiple sclerosis (MS). Magnetic resonance imaging shows the encephalomyelitic plaque of the pontomedullary region (*arrows*) involving either the entry zone of the eighth nerve or the vestibular nuclei. In both patients the plaque mimicked a peripheral vestibular lesion with contraversive rotational spontaneous nystagmus, ipsiversive direction of fall, nausea and emesis as well as an ipsilateral hyporesponsiveness of the horizontal semicircular canal to caloric irrigation

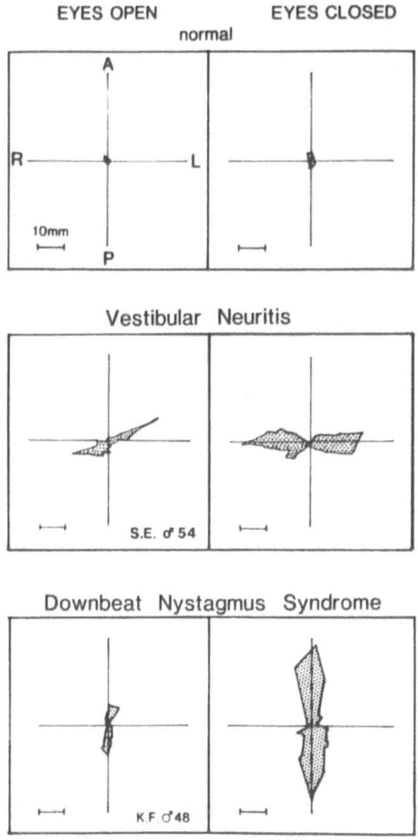

Fig. 2. Pathologic lateral body sway in vestibular neuritis and predominant fore-aft body sway in downbeat nystagmus syndrome. Sway histograms for fore-aft (*A*, *P*) and lateral (*R*, *L*) postural sway during upright stance with eyes open (*left*) and eyes closed (*right*) obtained with a force-measuring platform. For comparison, see registration of body sway in a normal subject (*top*). Preferred direction of postural instability and body sway is in the lateral direction for acute vestibular neuritis (*center*) but fore-aft in downbeat nystagmus syndrome (*bottom*). Body sway increases significantly with eyes closed when visual stabilization cannot compensate for the peripheral (*center*) or central (*bottom*) vestibular tone imbalance in either the yaw plane (vestibular neuritis) or pitch plane (downbeat nystagmus syndrome)

of the fourth ventricle could be the area responsible, because structures serving both vestibular and trigeminal motor function are present therein. A demyelinating plaque in multiple sclerosis may also mimic signs of vestibular neuritis (Fig. 1), including caloric hyporesponsiveness and a tendency to fall away from the side of the lesion (Fig. 2). Lesions are then usually located at the root entry zone of the eighth nerve [6] or the vestibular nuclei [15]. These brain-stem disorders may involve more than the horizontal semicircular canal pathways, and there are many cases that do not confirm to this classification scheme.

Downbeat Nystagmus/Vertigo Syndrome

Downbeat nystagmus in the primary position of gaze, accompanied by oscillopsia, is a clearly defined syndrome almost specific to structural lesions of the paramedian craniocervical junction. This is not purely an oculomotor disorder, but a central vestibular disorder that affects perception and balance.

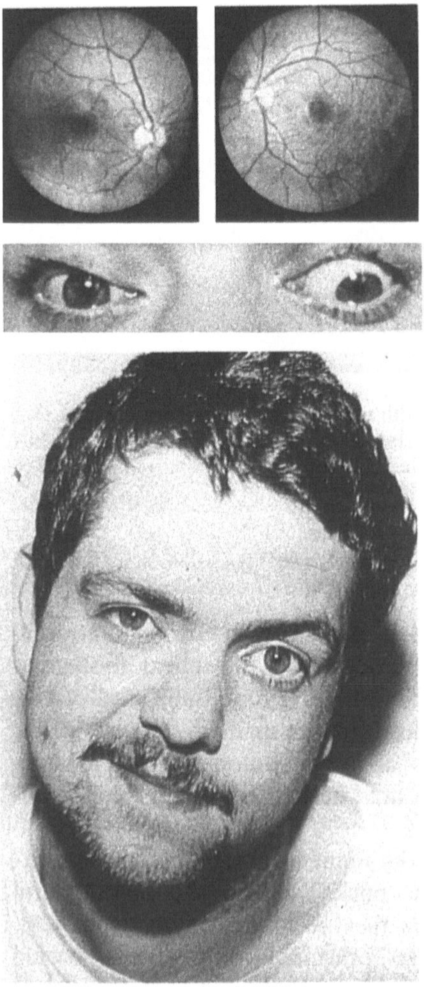

Fig. 5. Tonic OTR to the left in a patient suffering from right paramedian mesodiencephalic hemorrhage including the INC. Fundus photographs show conjugate ocular torsion clockwise from the viewpoint of the observer (*top*); the eyes show vertical divergence (skew deviation, *middle*); the head is tilted clockwise (*bottom*)

Ocular Tilt Reaction (OTR)

This eye-head synkinesis consists of lateral head tilt, skew deviation of the eyes (hypotropia of the undermost eye), and ocular torsion (Fig. 5) as first clearly delineated by electrical stimulation of the rostral midbrain tegmentum of the monkey [26]. It represents a fundamental pattern of coordinated eye-head roll motion based upon both utricular and vertical canal input [4]. In clinical cases of OTR, structural or functional lesions involve graviceptive pathways and the pathways for the VOR in roll from the medulla oblongata [4] to mesodiencephalon [18], including the interstitial nucleus of Cajal (INC) and the rostral interstitial nucleus of the medial longitudinal fascicle. OTR is not a rare clinical syndrome but occurs in about 30% of cases as a feature of an acute Wallenberg's syndrome [11] and in about 20% of acute

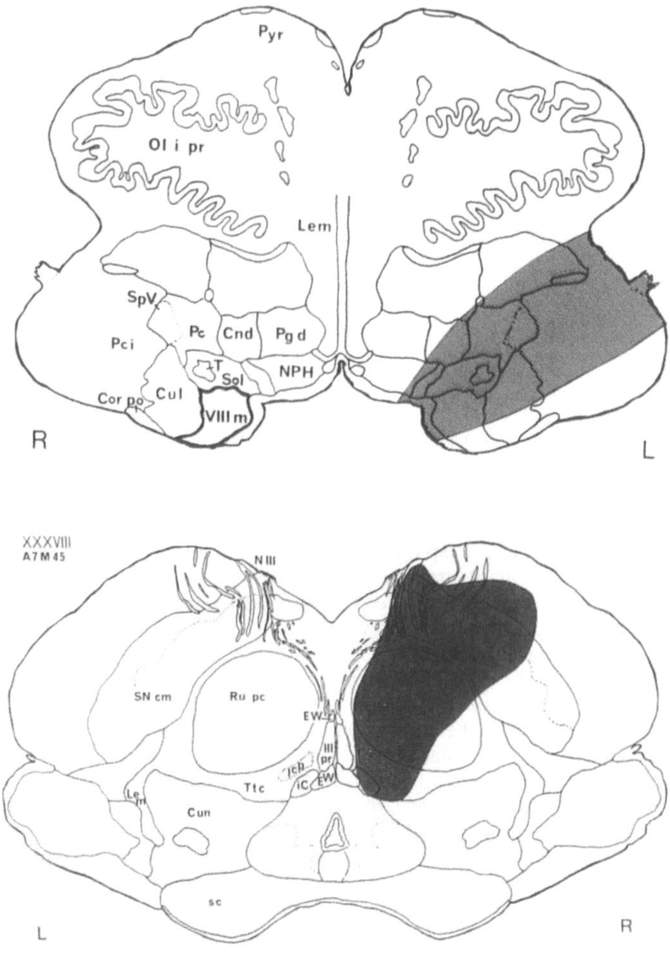

Fig. 6. Typical vestibular lesions in OTR, caused by lateral medullary infarctions (Wallenberg's syndrome, *top*) or paramedian thalamic infarctions including the INC in the rostral midbrain (*bottom*). When the lesion of the MRI scan is projected onto the transverse section of the medulla oblongata (*top*) or the rostral midbrain (*bottom*) (as adapted from [23]), involvement of the vestibular nuclei and the INC can be demonstrated

unilateral brain-stem infarctions [5]. All unilateral brain-stem lesions caudal to the upper pons cause ipsiversive tilts in the roll plane, whereas all lesions rostral to this pontine level cause contraversive tilts. This was convincingly demonstrated by correlation of the data with CT and MRI findings and the clinical syndrome (Fig. 6). Otolithic and vertical semicircular canal inputs to the interstitial nucleus of Cajal from the contralateral vestibular nucleus could be involved in the pathological tilt of head, eye, and perception. Contraversive tilts due to unilateral mesencephalic lesions (as opposed to the ipsiversive tilts in pontomedullary lesions) suggest the existence of a

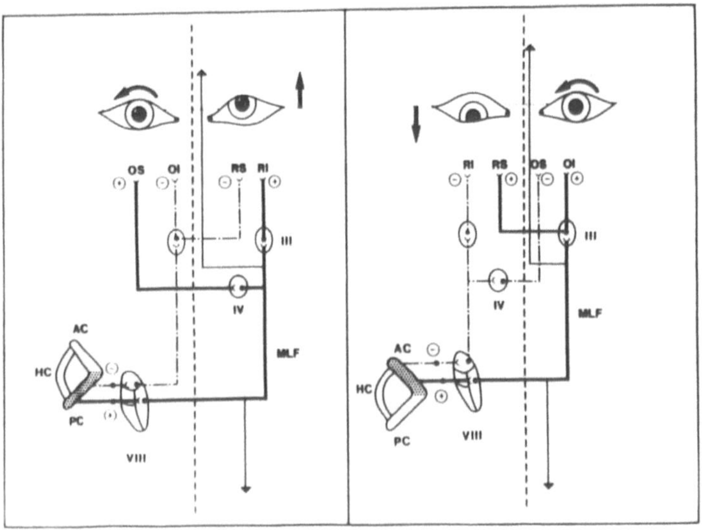

Fig. 7. Hypothetical explanation of OTR due to lesions of the vertical semicircular canal pathways. Schematic drawing of the three-neuron vestibulo-ocular reflex arc between the posterior (*left*) and the anterior (*right*) semicircular canal and the extraocular eye muscles. An excitatory ascending pathway is linked from the posterior semicircular canal to the ipsilateral superior oblique and the contralateral inferior rectus muscle (*left*). An inhibitory ascending pathway is linked to the ipsilateral inferior oblique and the contralateral superior rectus muscle. A lesion of these pathways causes excyclotropia of the ipsilateral and hypertropia of the contralateral eye. An excitatory ascending pathway projects from the anterior semicircular canal to the ipsilateral superior rectus and the contralaterial inferior oblique muscles (*right*). An inhibitory pathway projects to the ipsilateral inferior rectus and the contralateral superior oblique muscle. A lesion of these pathways of the anterior semicircular canal would cause a hypotropia of the ipsilateral and incyclotropia of the contralateral eye. A combination of both anterior and posterior semicircular canal pathways induces a complete OTR consisting of hypotropia of the ipsilateral eye, hypertropia of the contralateral eye, and conjugate ocular torsion of both eyes toward the undermost eye in the direction of head tilt. *AC*, *PC*, and *HC*, anterior, posterior, and horizontal semicircular canal; *MLF*, medial longitudinal fasciculus; *OS* and *OI*, superior oblique and inferior oblique muscles; *RS* and *RI*, mm. rectus superior and inferior

crossed "VOR pathway" between the vestibular nucleus and the contralateral INC.

The pattern of OTR in patients with Wallenberg's syndrome differs from "midbrain OTR" in that ocular torsion is disconjugate (or monocular) with predominant excyclotropia of the ipsilateral hypotropic eye (Fig. 7). Thus, we hypothesize a lesion of the vertical semicircular canal pathways. A combined lesion of the vertical semicircular canal and otolithic pathways is possible, because they travel closely together from the vestibular nuclei (mainly the rostral medial vestibular nuclei) toward the contralateral medial longitudinal fascicle. We have reason to believe that different patterns of OTR in patients with Wallenberg's syndrome reflect different combinations of anterior and posterior semicircular canal and otolith dysfunction [11].

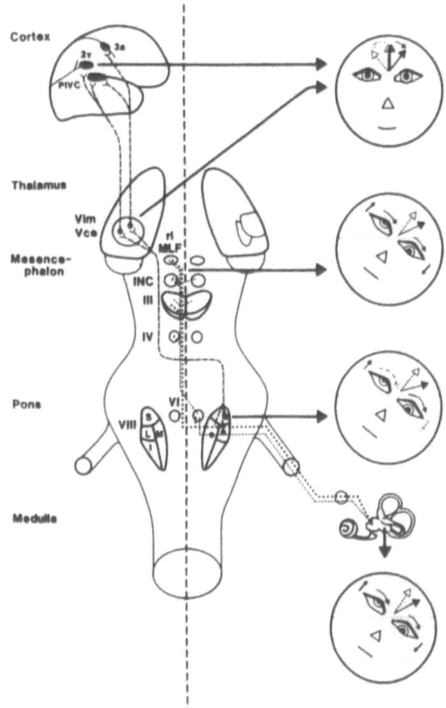

Fig. 8. Graviceptive pathways mediating vestibular reactions in the roll plane. The projections from the otoliths and vertical semicircular canals to the ocular motor nuclei (nucleus trochlearis IV, nucleus oculomotorius III, nucleus abducens VI) and the supranuclear centers of the INC, and the rostral interstitial nucleus of the medial longitudinal fasciculus (*riMLF*) are shown. They subserve VOR in three planes. The VOR is part of a more complex vestibular reaction which also involves vestibulospinal connections via the medial and lateral vestibulospinal tracts for head and body posture control. Furthermore, connections to the vestibular cortex (areas *2v* and *3a* and the parietoinsular vestibular cortex) via the vestibular nuclei of the thalamus (*Vim*, *Vic*) are depicted. Graviceptive vestibular pathways for the roll plane cross at pontine level. OTR is depicted schematically on the right in relation to the site of the lesion: ipsiversive OTR with peripheral and pontomedullary lesions; contraversive OTR with pontomesencephalic lesions. In vestibular thalamus and cortex lesions the tilts of SVV may be contraversive or ipsiversive; OTR is not induced by supratentorial lesions above the level of INC

Deviation of subjective visual vertical (SVV), ocular torsion of the eyes, and lateropulsion of the body are the perceptual, ocular motor, and postural consequences of a common lesion of central vestibular pathways that subserve the vestibulo-ocular reflex in the roll plane.

Deviations of the position of the eye in the roll plane (ocular torsion) and of the SVV were systematically studied in 111 patients with acute unilateral vascular brain-stem lesions. Ninety-four percent showed a direction-specific tilt of the SVV and 83% exhibited a pathologic ocular torsion of one or both eyes. Ocular torsion and SVV tilts are therefore among the most sensitive signs in acute unilateral brain-stem dysfunction [5, 12]. A complete OTR or its single features, such as ocular torsion, skew deviation, SVV tilt, and body lateropulsion, may all reflect dysfunction of the VOR in the roll plane (Fig. 8).

The presented classification of central vestibular brain-stem disorders according to VOR planes in its present form is speculative, since it combines semicircular canal and otolith function. It is incomplete, since other vestibular syndromes such as infratentorial (central-) positional nystagmus/vertigo and supratentorial syndromes such as thalamic astasia and vestibular epilepsy are not yet included. Our aim was to establish a platform from which the variety of central vestibular syndromes can be logically ordered with respect to anatomical structures and function.

References

1. Anastasio TJ, Robinson DA (1990) Distributed parallel processing in the vertical vestibulo-ocular reflex: learning network compared to tensor theory. Biol Cybern 63:161–167
2. Baloh RW, Spooner JW (1981) Downbeat nystagmus. A type of central vestibular nystagmus. Neurology (NY) 31:304–310
3. Brandt T (1991) Man in motion. Historical and clinical aspects of vestibular function. Brain 114:2159–2174
4. Brandt T, Dieterich M (1987) Pathological eye-head coordination in roll: tonic ocular tilt reaction in mesencephalic and medullary lesions. Brain 110:649–666
5. Brandt T, Dieterich M (1992) Cyclorotation of the eyes and subjective visual vertical in vestibular brain stem lesions. Ann NY Acad Sci 656:537–549
6. Brandt T, Dieterich M, Büchele W (1986) Postural abnormalities in central vestibular brain-stem lesions. In: Bles W, Brandt T (eds) Disorders of posture and gait. Elsevier, Amsterdam, pp 141–156
7. Büchele W, Brandt T (1988) Vestibular neuritis, a horizontal semicircular canal paresis? Adv Otorhinolaryngol 42:157–161
8. Büchele W, Brandt T, Degner D (1983) Ataxia and oscillopsia in downbeat nystagmus/vertigo syndrome. Adv Otorhinolaryngol 30:291–297
9. Carpenter MB, Cowie RJ (1985) Connections and oculomotor projections of the superior vestibular nucleus and cell group "Y". Brain Res 336:265–287
10. De Jong IMBV, Cohen B, Matsuo V, Uemura T (1980) Midsagittal ponto-medullary brain stem section: effects of ocular adduction and nystagmus. Exp Neurol 68:420–442
11. Dieterich M, Brandt T (1992) Wallenberg's syndrome: lateropulsion, cyclorotation, and subjective visual vertical in thirty-six patients. Ann Neurol 31:399–408
12. Dieterich M, Brandt T (1993) Ocular torsion and tilt of subjective visual vertical are sensitive brain stem signs. Ann Neurol 33:292–299
13. Duncan GW, Parker SW, Fisher CM (1975) Acute cerebellar infarction of the PICA territory. Arch Neurol 32:364–368
14. Fisher A, Gresty M, Chambers B, Rudge P (1983) Primary position upbeating nystagmus: a variety of central positional nystagmus. Brain 106:949–964
15. Francis DA, Bronstein AM, Rudge P, DuBoulay EPGH (1992) The site of brain-stem lesions causing canal paresis: an MRI study. J Neurol Neurosurg Psychiatry 55:446–449

16. Gilmann S, Baloh RW, Tomyasu U (1977) Primary positional upbeat nystagmus. Neurology 27:294–298
17. Graf W, Ezure K (1986) Morphology of vertical canal-related second-order vestibular neurons in the cat. Exp Brain Res 63:35–48
18. Halmagyi GM, Brandt T, Dieterich M, Curthoys IS, Stark RJ, Hoyt WF (1990) Ocular tilt reaction due to unilateral meso-diencephalic lesion. Neurology 40:1503–1509
19. Hopf HC (1987) Vertigo and masseter paresis. A new local brain-stem syndrome probably of vascular origin. J Neurol 235:42–45
20. Ito M, Nisimaru N, Yamamoto M (1976) Pathways for the vestibulo-ocular reflex excitation arising from semicircular canals of rabbits. Exp Brain Res 24:257–271
21. Keane JR, Itabashi HH (1987) Upbeat nystagmus: clinicopathologic study of two patients. Neurology 37:491–494
22. Leigh RJ, Brandt T (1993) A re-evaluation of the vestibulo-ocular reflex: new ideas of its purpose, properties, neural substrate, and disorders. Neurology (submitted)
23. Olszewski J, Baxter D (1982) Cytoarchitecture of the human brain stem, 2nd edn. Karger, Basel
24. Ranally RJ, Sharpe JA (1988) Upbeat nystagmus and the ventral tegmental pathway of the upward vestibulo-ocular reflex. Neurology 38:1329–1330
25. Rubinstein RL, Norman DM, Schindler RA, Kaseff L (1980) Cerebellar infarction. A presentation of vertigo. Laryngoscope 90:505–514
26. Westheimer G, Blair SM (1975) The ocular tilt reaction – a brain-stem ocular motor routine. Invest Ophthalmol 14:833–839
27. Zee DS, Yamazaki A, Butler PHZ, Gücer F (1981) Effect of ablation of flocculus and paraflocculus on eye movements in primate. J Neurophysiol 46:878–899

Elektronystagmography: The Answers One Might Get

V. Henn

Neurologische Abteilung, Universitätsspital, Rämistraße 100, 8091 Zürich, Switzerland

Introduction

Electronystagmography might be a very useful instrument to reveal pathology of eye movements and of the vestibular system that cannot be detected at the bedside; often, it is very useful to have documentation and quantitative measurements. On the other hand, if used uncritically, results might be very misleading, e.g., if the patient did not understand instructions or did not follow them, or if inadequate stimuli or recording devices were used. The original procedure as published by Jung and Kornhuber [15] is still to be recommended. Brandt and Büchele [6] discuss further details of application, while Leigh and Zee [16], Baloh and Honrubia [3], and Brandt [5] review recent clinical literature. If the following is very critical concerning the choice of stimuli and interpretation of results, I also want to stress that for many situations electronystagmography is still a very valuable method which cannot easily be replaced, as it tests the functional integrity of several brainstem systems.

Eye Movements Occur in Three Dimensions, While Electronystagmography Can Measure Only Two Dimensions

Skin electrodes are attached to the outer canthi of the eyes for horizontal, and above and below one eye for vertical recording. Torsional movements cannot be recorded with this method, and vertical eye positions are difficult to calibrate as lid movements interfere with the signals. The retino-corneal potential changes up to 50% with light adaptation [1]. If changing light conditions exceed several minutes, as during testing of vestibular reflexes in the dark, several repetitions of the calibration are necessary for quantitative analysis.

Amplifiers are usually set to 0.1–30 Hz band width. With dc recording one inevitably gets some baseline drifting. Optimal are amplifiers that offer the option to choose between dc and different time constants, or that have the option of an automatic dc reset, if drift exceeds a certain voltage.

L.R. Caplan and H.C. Hopf (Eds.)
Brain-Stem Localization and Function

The upper frequency limit is conveniently set to 30 Hz to avoid 50- or 60-Hz noise. If one wants to determine velocity and acceleration profiles of saccades, a band width of 0–300 Hz is mandatory, as rapid eye movements have frequency components well beyond 30 Hz [2].

Spontaneous Nystagmus: Possible Artifacts and Their Detection

With electronystagmography, the limit of resolution is on the order of 0.5 deg. Sometimes it might be difficult to recognize spontaneous nystagmus in darkness because of continuous lid artifacts or other artifacts caused by periorbital muscle activation. If spontaneous nystagmus is suspected, it should always be verified by independent means [12]. The easiest way is to ask the patient to look straight ahead, to cover the contralateral eye, and then to inspect the papilla with an ophthalmoscope. As the papilla has a diameter of 5 deg, it is easy to estimate the velocity of nystagmus. For instance, if the eye drifts half a papilla diameter during 1 s, the velocity can be judged to be 2–3 deg/s. As one looks at the inner surface of the eye, the direction is inverted. This is a very sensitive method for detecting any spontaneous nystagmus. Another elegant method is to inspect the eyes in total darkness with an infrared viewing device. The third possibility is to inspect the patient while he or she is wearing Frenzel glasses. With most such glasses, some light reflexes are seen which might partially suppress nystagmus. However, most authors agree that spontaneous nystagmus must exceed a certain velocity, e.g., 5 deg/s in total darkness, to be of pathological significance, and such level of nystagmus, even if reduced with Frenzel glasses, can always be detected with one of the above-described methods.

Saccadic Pursuit

To test pursuit movements requires that the patient clearly understands instructions and does what is required. If one uses a pendulum with a simple marking, or a moving light point for pursuit, the patient might be easily distracted, e.g., by the presence of the technician on one side. We obtained improved pursuit when we attached a small one-digit display to the moving pendulum. The display shows numbers and letters in random order, and we tell the patient that his task is to identify these numbers as quickly as possible and remember them for later recall. Without being told that they should pursue, most patients do better in this way than with any other instruction we have tried. The ability to pursue without catch-up saccades is a complex function of stimulus velocity, acceleration, and exposure time. In addition, it is an age-dependent ability. While a unilateral saccadic pursuit is always pathological, the minimum velocity one should be able to pursue is usually given as 30 deg/s.

Optokinetic Nystagmus

Investigations using a large number of normal subjects show that the gain of optokinetic nystagmus progressively declines above 40 deg/s stimulus velocity, and that with higher stimulus velocities responses show an increasingly larger standard deviation. It was therefore recommended that optokinetic stimulation be confined to low velocities. Although we do not dispute the experimental finding, we do not agree with the suggested consequence. Our task is to detect possible pathology in the single patient. If we apply high-velocity optokinetic stimulation, e.g., 100 deg/s, a side difference might be significant, and nystagmus velocity should be clearly higher than with 40 or 60 deg/s stimulus velocity. Although standard deviations might be high with such velocities, our clinical task is to detect pathology in the single patient, not to set boundaries for responses between populations of patients and of normal subjects. With such a task in mind, it might often be useful to test a functional system at the limits of its performance, for instance, during high-velocity optokinetic nystagmus, where subtle side differences might manifest themselves, even if responses can show quite a high variability related to a normal population.

Another factor might bring out pathology during optokinetic stimulation: although one obtains the most reliable response when the patient sits inside a moving drum, small differences in maximum velocity between right and left might be better revealed by using a small hand-held drum rotated in front of the patient [10].

For normal elicitation of optokinetic nystagmus, two processes are responsible. One responds to large moving fields, especially in the periphery of the visual field; the other relies on foveal and parafoveal input and elicits pursuit movements. The two systems overlap widely in terms of retinal input [7] and have been called indirect or optokinetic, and direct or pursuit systems [8, 18, 21]. The pursuit system responds to higher velocities and seems to be more vulnerable. Therefore, with a small hand-held drum one might have a better chance of detecting pathology. The indirect system responds with a lower maximum velocity, but its response is more robust and reflex-like.

Vestibular Nystagmus

Conveniently, vestibular nystagmus is elicited by accelerating the patient to a velocity of 100 deg/s, rotating him with constant velocity over 30 s, and then decelerating. Two measures characterize the response: the gain and the time constant. Gain is defined as the ratio between nystagmus and stimulation velocity. Its value for horizontal nystagmus is usually between 0.6 and 0.8. The time constant indicates the time elapsed between maximum response and its decline to 37%. Although the decline is not strictly ex-

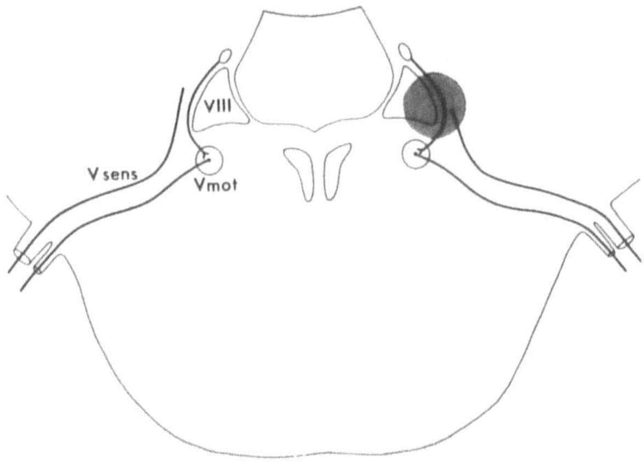

Fig. 1. Section through the brain-stem at the trigeminal motor nucleus level. *V sens*, sensory portion of the trigeminal nerve; *V mot*, trigeminal motor nucleus; *VIII*, vestibular nuclei; *shaded area*, represents lesion involving vestibular nuclei and the masseter reflex arc

lesions of the area in question. Far more frequently observed are patients with vertigo showing paresis of the lateral semicircular canal and impairment of the MassR. One such patient has been reported previously ([4], patient 3). Since then we have observed 46 similar patients with obvious vascular pathology (unpublished). Clinical improvement always coincided with improvement of MassR latencies and amplitudes, indicating that both abnormalities were caused by the same lesion.

In summary, the occurrence of acute peripheral-type vestibular disorder of intra-axial origin is now well established. Vertigo and tinnitus may be the only symptoms [6]. Clinically, this condition cannot be differentiated from "apoplexia vestibuli" or "vestibular neuronitis". Electrophysiologically, the characteristic features are the milder degree of canal paresis with caloric stimulation and the lesser effect of removal of fixation on the nystagmus [2, 7], as well as impairment of the masseter reflex [4], the latter being the most reliable sign.

References

1. Baloh RW, Honrubia V (1979) Clinical neurophysiology of the vestibular system. Davis, Philadelphia
2. Francis DA, Bronstein AM, Rudge P, du Boulay EPGH (1992) The site of brainstem lesions causing semicircular canal paresis: an MRI study. J Neurol Neurosurg Psychiatry 55:446–449
3. Hallpike CS, Harrison MS, Slater E (1951) Abnormalities of the caloric rest results in certain varieties of mental disorder. Acta Otolaryngol (Stockh) 39:151–159
4. Hopf HC (1987) Vertigo and masseter paresis. A new brainstem syndrome. J Neurol 235:42–45

5. Nieuwenhuys R, Voogd J, van Huijzen C (1988) The human central nervous system, 3rd edn. Springer, Berlin Heidelberg New York, pp 130–131
6. Tettenborn B, Dieringer H, Maurer J, Hopf HC, Krämer G, Thömke F (1992) Vertigo and tinnitus as the only indicator of brainstem disease. Neurology 42 [Suppl 3]:249
7. Uemura T, Cohen B (1973) Effects of vestibular nuclear lesions on vestibulo-ocular reflexes and posture in monkeys. Acta Otolaryngol (Stockh) [Suppl 256]:1–66

Oculomotor Syndromes in Rostral Brain-Stem Lesions

D. Kömpf

Klinik für Neurologie, Universität Lübeck, Ratzeburger Allee 160, 23562 Lübeck, Germany

Introduction

Eye movements serve vision by placing the image of an object of regard on the fovea of each retina and by preventing slippage of images on the retina. The brain has two modes of oculomotor control: Fast eye movements (saccades) and smooth (slow)-pursuit eye movements (SPEM). Fast eye movements bring the fovea to targets that fall on the extrafoveal retina, and SPEM maintain fixation of slowly moving targets. Six distinct oculomotor systems (saccades, smooth pursuit, VOR, vergence, optokinetics, and fixation) are utilized to fulfill the task. This brief survey will concentrate mainly on saccadic and SPEM disturbances in rostral brain-stem pathology. A review of basic anatomy and physiology of the oculomotor system will allow a better understanding of the pathogenetic mechanisms underlying the complex oculomotor brain-stem syndromes.

The saccadic system (Fig. 1) arises mainly from the frontal eye field (FEF, area 8). The FEF projects directly to the pontine saccade-generating center (PPRF); this pathway decussates at the junction of the midbrain and the pons. Saccades from the occipital lobe are abolished by ipsilateral superior colliculus (SC) lesions and are unaffected by lesions of the FEF or parietal area 7, indicating that occipital saccades have direct access to the brain stem via the SC. There is another indirect pathway through the caudate nucleus, the pars reticulata of the substantia nigra, and the SC; this pathway is tonically inhibitory to the SC.

From the PPRF – the *horizontal* saccade-generating center – axons project to the ipsilateral abducens nucleus containing both neurons to the ipsilateral lateral rectus muscle and smaller interneurons that project – after decussation – in the contralateral medial longitudinal fasciculus (MLF) to the contralateral medial rectus subnucleus.

The innervational change during a saccade consists of a phasic discharge (pulse, eye velocity command) and a tonic discharge (step, eye position command). The phasic discharge consists of a high-frequency pulse of innervation that moves the eyes rapidly against orbital viscous forces. When the new position is attained, the position command is required to sustain eye

L.R. Caplan and H.C. Hopf (Eds.)
Brain-Stem Localization and Function

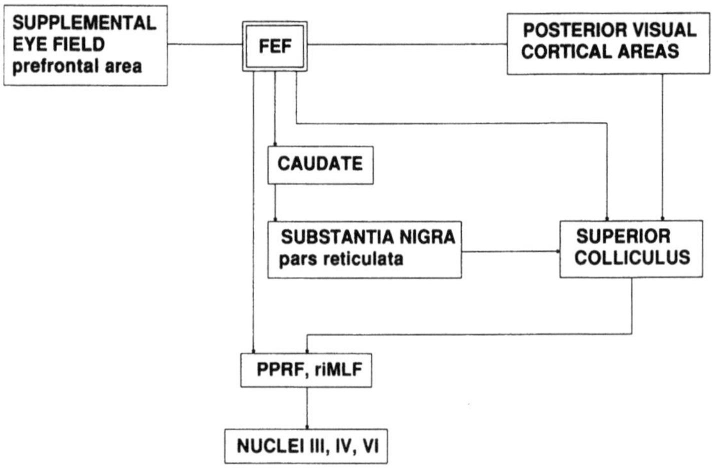

Fig. 1. Major structures that participate in the control of saccades. *FEF*, frontal eye field; *PPRF*, paramedian pontine reticular formation; *riMLF*, rostral interstitial nucleus of medial longitudinal fasciculus; *III*, *IV*, *VI*, oculomotor, trochlear, and abducens nerves

position against the elastic restoring forces of the eye muscles. The pulse is generated by burst neurons in the PPRF. Between saccades the burst neurons are kept silent by pause neurons located in the midline of the caudal pons (n. raphe interpositus). An inhibition of the pause neurons must release the burst neurons to create the pulse. The step discharge, the position command, is generated from the pulse, the velocity command by a neural integrator that "integrates" (in the mathematical sense) the pulse. This neural integrator is located in the medial vestibular neurons and the nucleus prepositus hypoglossi, which lies just medial to the vestibular nuclei [13].

The *vertical* saccadic-generating center is situated in the rostral midbrain rostromedial to the red nucleus. It has been delineated by stimulation and lesion studies [11], single-cell recordings [2], and anatomic tracer studies [3] and was termed by Jean Büttner-Ennever the rostral interstitial nucleus of the MLF (riMLF), about 15 years ago. The riMLF receives input from the PPRF and also directly from the FEF. As a general principle, vertical gaze is mediated by bilateral circuits.

Even though the up/down question is not yet solved completely, there is much experimental and clinical evidence indicating that downgaze seems to be situated in the lateral aspect of the riMLF, projections going down from here on both sides – uncrossed – to the cranial nerve nuclei III and IV (Fig. 2). With upgaze the projection of the medial part of the riMLF is not a direct projection, however; upgaze riMLF burst neurons project across the posterior commissure (PC), then travel down dorsally close to the midline to motor neurons in the third nerve nuclei.

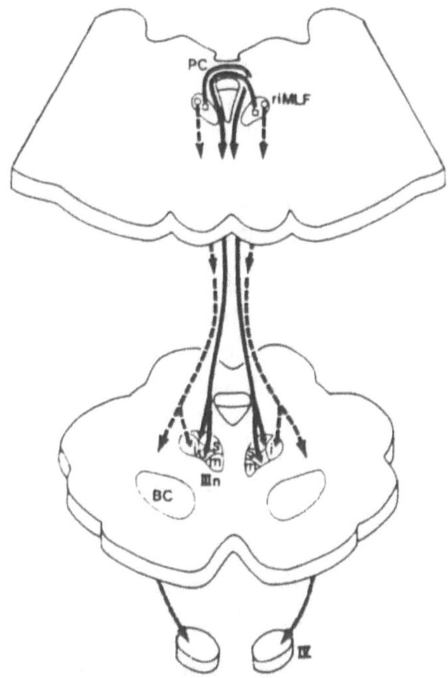

Fig. 2. Structures involved in the control of vertical saccades. *BC*, brachium conjunctivum; *PC*, posterior commissure; *riMLF*, rostral interstitial nucleus of medial longitudinal fasciculus; *IIIn*, oculomotor nucleus; □ –, burst cells for upward saccades; o---, burst cells for downward saccades

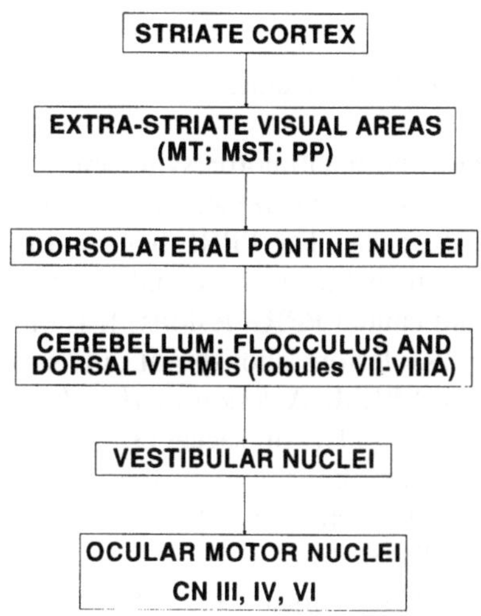

Fig. 3. Putative pursuit pathway. *MT*, middle temporal visual area; *MST*, mediosuperior temporal area; *PP*, posterior parietal region; *CN*, cranial nerves

The SPEM system starts where vision is primarily received (area 17) responding to slippage of an image near the fovea. Axons from the striate cortex project to extrastriate visual areas – middle temporal visual area (MT), mediosuperior temporal area (MST), posterior parietal (PP) region, from here ipsilaterally to the dorsolateral pontine nuclei, then to the flocculus and the dorsal vermis of the cerebellum, then to the oculomotor nuclei. These structures are part of an ipsilateral parieto-occipito-ponto-cerebellar pathway (Fig. 3). It is interesting that this pathway, unlike the saccadic pathway, goes from the brain stem to the cerebellum and then back to the brain stem, whereas the saccadic pathway goes directly from the cerebral hemisphere to the brain stem [9, 13].

Eye movement abnormalities that are due to brain-stem disorders include:

- Gaze palsy (both SPEM low-gain disorder and Saccades)
- Internuclear ophthalmoplegia (INO)
- WEBINO (walleyed bilateral INO)
- One-and-a-half syndrome
- Optomotor nerve palsies III, IV, VI (nuclear, fascicular syndromes)
- Nystagmus phenomena
- Ocular bobbing, dipping
- Locked-in syndrome
- Ocular tilt reaction, skew deviation
- Vergence disturbances
- VOR abnormalities

A gaze palsy can be horizontal or vertical; the paresis can abolish all types of eye movements in this plane or dissociatively disturb single subsystems of eye movements only.

Pontine Syndromes

Pontine oculomotor syndromes are characterized by abnormalities of horizontal eye movements. A *PPRF lesion* produces a permanent ipsilateral gaze palsy. The PPRF begins at the rostral portion of the pons and stretches down to the most caudal portion of the pons, so that there can be rostral and caudal PPRF lesions. It is only at the level of the abducens nucleus that PPRF lesions abolish not only saccades but also vestibulo-ocular movements and SPEM. A *lesion of the abducens nucleus* always causes ipsilateral conjugate palsy of saccades, pursuit, and vestibular movements: the internuclear abducens neurons are the most important VOR pathway to the medial rectus, and they also transmit the saccadic and pursuit signals to the medial rectus.

MLF lesions cause an internuclear ophthalmoplegia (INO): impaired adduction on the side of a lesion in the MLF and abducting jerk nystagmus on the opposite side. Usually, patients with bilateral INO have their eyes in primary position; they are orthotropic, and only occasionally do they become exotropic ("walleyed").

The *one-and-a-half syndrome* is the result of a combined PPRF-MLF lesion, in the acute stage exhibiting "paralytic pontine exotropia": the opposite eye (opposite to the side of the lesion) is abducted, exotropic.

Table 1. Pathology of saccades in brain-stem disorders

Disorders of velocity
slow saccades
Disorders of accuracy
dysmetria
(hypermetria, lateropulsion, hypometria)
Disorders of initiation
ocular motor apraxia
Inappropriate saccades (intrusions and oscillations)
square wave jerks
macro square wave jerks
macro saccadic oscillations
ocular flutter
opsoclonus

Skew deviation refers to vertical divergence of the eyes caused by supranuclear lesions. Skew is attributed to disruption of projections from the utricles that maintain tonic innervation of vertically acting oculo-motor neurons. The lower skewed eye is usually on the side of the lesion in the pons (midbrain or medulla). In unilateral INO skewing may occur with the higher eye on the side of the MLF lesion. For different types of skew deviation and ocular tilt reaction see the chapters by Brandt and Dieterich, in this volume.

The innervational change during all types of saccadic eye movements (pulse/step command) can be disturbed in different ways so that varied abnormalities of saccadic trajectory can occur (Table 1). A disturbance of the pulse generator in the PPRF with decrease in the height of the pulse causes a *general slowing* of saccadic movements with a wide range of differential diagnoses, including general oculomotor apraxia, Huntington's disease, Wilson's disease, OPCA, progressive supranuclear palsy, Whipple's disease, lipid storage disease, anticonvulsants toxicity, PPRF lesions, and peripheral processes (myasthenia grais, chronic progressive external ophthalmoplegia, thyroid disease). The pulse and step must be appropriately matched; a mismatch creates *postsaccadic drift or glissades*. If the neural integrator in the nucleus prepositus hypoglossi is disturbed, the saccadic step cannot be sustained; the eyes drift back towards primary position, creating *gaze-evoked nystagmus*. A *hypometric saccade* results if the pulse amplitude is too small but pulse and step are matched appropriately. Disturbances of pause neurons in the nucleus raphe interpositus can cause different forms of *ocular oscillations* without intersaccadic intervals, like ocular flutter or opsoclonus.

Ocular bobbing consists of intermittent conjugate rapid downward movement of the eyes, followed by a slower return to the primary position [15]. The movements are classically seen with intrinsic pontine lesions, but they have been encountered with posterior fossa lesions that secondarily compress the pons. Horizontal eye movements are absent. An inverse form of

Table 2. Smooth-pursuit eye movement disorders in CNS disease

Disorder	CNS disease
Unilateral low-gain pursuit *ipsilateral* to the lesion	parietal lobe brain stem (mesencephalon and rostral pons) cerebellum
Bilateral low-gain pursuit with *bilateral* disease	bilateral cerebral " basal ganglia " brain stem " cerebellar normal aging inattention schizophrenia
Bilateral CNS disease with *normal* SPEM	(almost) none

bobbing consists of slow downward movements and a rapid return to mid-position (ocular dipping). Reverse bobbing consists of rapid deviation of the eyes upward and a slow return to the horizontal. These forms are less reliable for localization.

On the pontine level there is a clear anatomical and functional segregation of saccades and pursuit eye movements. In a pontine *lesion involving the dorsolateral pontine nuclei* an isolated ipsilateral low-gain pursuit manifested by ipsilateral saccadic pursuit can result [16] – as in unilateral disease of the parietal lobe or the cerebellum. With bilateral brain-stem disease there will be a bilateral pursuit abnormality – a rather nonspecific finding which can be registered in almost any kind of CNS disease, even in schizophrenia or normal aging (Table 2).

Midbrain Syndromes

The most rostral neuro-ophthalmological brain-stem signs are summarized in Table 3; the most characteristic sign consists of vertical gaze abnormalities. As a general rule, vertical gaze palsies signify bilateral or paramedian lesions.

An isolated *downgaze impairment* is a rare clinical event caused by a bilateral lateral riMLF lesion, almost always of vascular origin – syndrome of the posterior thalamo-subthalamic paramedian artery. Figure 4 shows the butterfly-shaped bilateral medial-thalamic lesion – the mesodiencephalic riMLF lesion not demonstrated – in a patient with a predominant downgaze palsy (vertical one-and-a-half syndrome) clinically accompanied by thalamic dementia [10]. This combination of oculomotor and behavioral signs is the hallmark of the "top of the basilar" syndrome first described by Caplan [4].

Table 3. Neuro-ophthalmological midbrain signs

Vertical gaze
Upgaze paresis
Downgaze paresis
Combined up/downgaze paresis
Sustained downgaze (in upgaze paresis)
Sustained upgaze (in downgaze paresis)
Vertical one-and-a-half-syndrome
Monocular upgaze paresis
Skew deviation
Pseudoabducens paralysis
Ocular tilt reaction
Convergence
Paralysis, spasm, convergence-retraction nystagmus
Accommodation
Spastic-paretic accommodation
Eye lids
Eye lid retraction (Collier's sign)
Palpebral nystagmus (Pick)
Oculomotor (unilateral/bilateral)
Third nerve palsy (supra/prenuclear, nuclear, fascicular)
Pseudo-Parinaud syndrome
Fourth nerve palsy (fascicular)
Internuclear ophthalmoplegia
Nystagmus
Upbeat nystagmus
See-saw nystagmus
Pupillary
Light-near dissociation
Inverse Argyll-Robertson
Corectopia
Peduncular hallucinosis

The high frequency of *upgaze palsies* can be explained by the crossed pathway for upward eye movements from the medial part of the riMLF, running down along the midline to reach the third nerve nucleus (Fig. 2). Therefore, a variety of locations of lesions in the midbrain will produce an upgaze palsy: PC lesions, bilateral pretectal lesions, and large tegmental lesions on one side or the midbrain tegmentum (Table 4). Although anatomically unilateral, lesions can interrupt pathways involved in vertical gaze just before and after they decussate, thus inducing functionally bilateral lesions. In the sample of 11 patients with different types of vertical gaze palsy due to unilateral midbrain lesions reported by Hommel and Bogousslavsky [7], patients had either upgaze palsy, combined up-downgaze palsy or a vertical one-and-a-half syndrome with bilateral upgaze palsy and ipsilateral or contralateral downgaze palsy caused by lesions of the PC and of fibers involved in downgaze. The reverse syndrome, bilateral downgaze palsy with unilateral elevation palsy, has been reported only twice, following infarction at the thalamo-mesencephalic junction [5, 10].

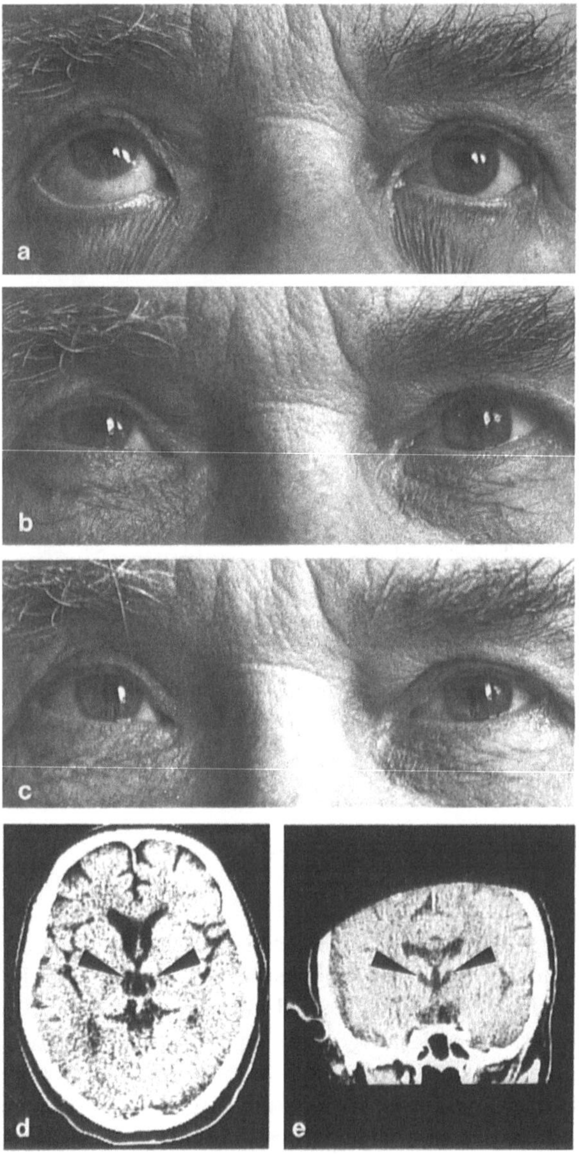

Fig. 4. **a** Predominant downgaze palsy (vertical one-and-a-half syndrome) and thalamic dementia due to a butterfly-shaped mediothalamic infarct. **a** Attempted upgaze with left elevator paresis; **b** primary position; **c** (attempted downgaze). Horizontal (**d**) and coronary (**e**) CT sections show syndrome of the posterior thalamo-subthalamic paramedian artery

Table 4. Up-/downgraze palsy

	Location	Etiology
Downgaze palsy	Mesodiencephalic (lateral riMLF)	Vascular (posterior thalamo-subthalamic paramedian artery)
Upgaze palsy	Pretectal (medial riMLF) posterior commissure, large midbrain tegmental lesion	Almost any pathological dorsal midbrain condition (TU, MS, vascular, encephalitis, aqueductal stenosis, uncal herniation)

Table 5. Parinaud's syndrome: manifestations and synonyms

Manifestations	Synonyms
Upgaze palsy (saccades and pursuit)	Pretectal syndrome
Convergence-retraction nystagmus	Posterior commissural syndrome
Pseudoabducens paralysis (convergence bias with horizontal saccades)	Sylvian aqueduct syndrome
Pupillary abnormalities (large pupils with light-near dissociation)	Dorsal midbrain syndrome
Corectopia (eccentric position of the pupils)	Kõrber-Salus-Elschnig syndrome
Spastic-paretic accommodation	Freund-Vogt Herdbildung
Pathologic lid retraction (Collier's sign)	

A distinct constellation of neuro-ophthalmological midbrain signs accompanies the characteristic upgaze deficit in Parinaud's syndrome, which has many eponyms and anatomic designations (Table 5).

Monocular elevation paresis is an infrequent manifestation of unilateral damage in the rostral midbrain signifying disruption of supranuclear projections, perhaps from the riMLF on one side, to contralateral superior rectus motor neurons. In most cases a pretectal lesion contralateral to the paretic eye has been blamed, but an ipsilateral tectal lesion may also be responsible for this double elevator palsy [6, 7].

Sustained *tonic ocular downward deviation* typically occurs in caudal thalamic hemorrhage affecting dorsal midbrain structures (Fig. 5, Table 6), often accompanied by convergence [8]. *Tonic upward deviation* however, occurs following severe hypoxic encephalopathy, sparing midbrain structures critical to vertical gaze.

Isolated *unilateral deficits of the superior colliculus* are rare in man, but quantitative electro-oculography in one case proved that this structure plays an important role both in the triggering and inhibition of reflexive visually

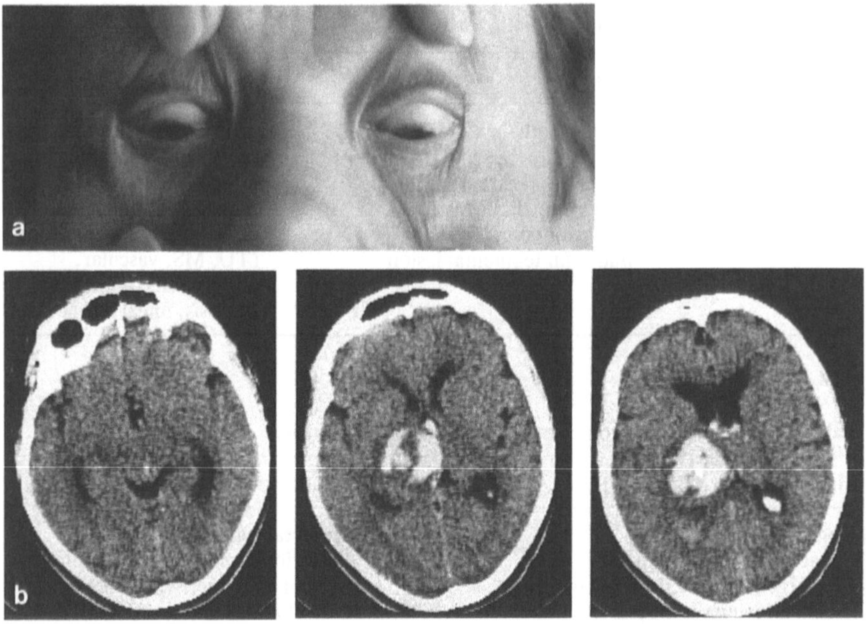

Fig. 5. a Tonic downgaze deviation in thalamic hemorrhage; **b** CT scan showing the hematoma of the thalamus involving the dorsal midbrain

Table 6. Sustained tonic vertical gaze deviation

Sustained downgaze	
Upgaze paresis	
Phenomenon of "wrong-way eyes"	
Small, sluggish or unreactive pupils	
Sensorimotor hemideficit	
State of responsiveness: obtunded to comatose	
Etiology:	Thalamic hemorrhage with subthalamic damage
	Hypoxia, Coma
Sustained upgaze	
Etiology:	Hypoxia
	Bilateral symmetric hemispheral and cerebellar subcortical damage *sparing* brain-stem structures

guided saccades. Increase of contralateral saccade latency, a high percentage of unwanted reflexive saccades, and contralateral hypometria are seen [14].

A unilateral *nuclear oculomotor complex infarction* is rare. This syndrome, which has been reported in altogether seven cases, manifests itself as ptosis when bilateral, as rectus superior paresis when contralateral, and as oculomotor (III) paresis (mostly incomplete) with or without pupillary

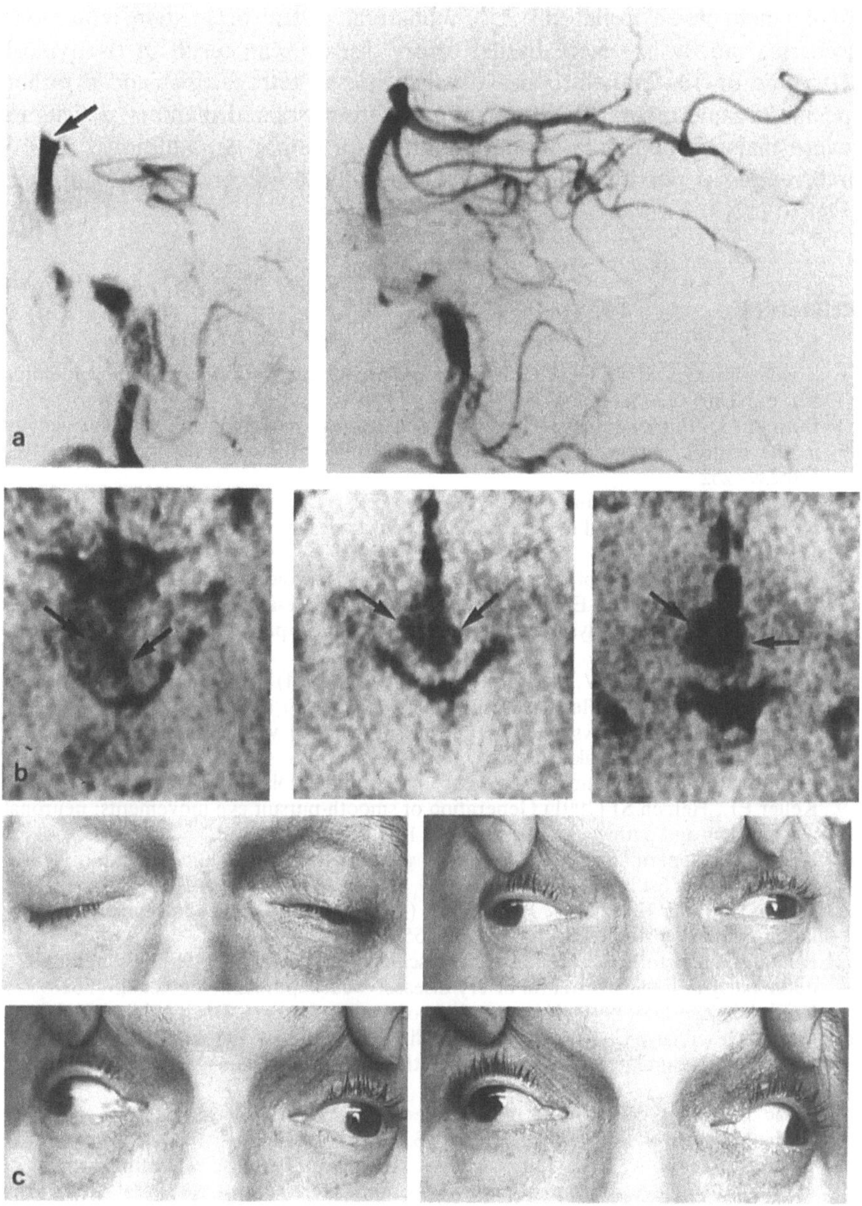

Fig. 6. **a** Basilar occlusion (*arrow*) at the superior cerebellar artery level (*left*) and recanalization after local intra-arterial fibrinolysis therapy (*right*); **b** CT scan showing bilateral tegmental mesencephalic hypodensity at different levels (*arrows*) and additional involvement of left medial thalamic nuclei (*right*); **c** bilateral third nerve palsy. *Above*, primary position; *lower left*, right gaze; *lower right*, left gaze

involvement when ipsilateral [1]. A bilateral rostral brain-stem infarction, occurring mostly in severe basilar artery disease, can result in the typical syndrome of bilateral third nerve palsy, spastic tetraparesis, and a rather specific organic mental syndrome ranging from isolated memory deficits to severe thalamic dementia, akinetic mutism, or coma. An additional riMLF involvement is possible and probable clinically, however, not recognizable (Fig. 6) [12].

References

1. Bogousslavsky J, Regli F (1983) Nuclear and prenuclear syndromes of the oculomotor nerve. Neuro-ophthalmology 3:211–216
2. Büttner U, Büttner-Ennever JA (1977) Vertical eye movement-related unit activity in the rostral mesencephalic reticular formation of the alert monkey. Brain Res 130:239–252
3. Büttner-Ennever JA, Büttner U (1978) A cell group associated with vertical eye movements in the rostral mesencephalic reticular formation of the monkey. Brain Res 151:31–47
4. Caplan R (1980) "Top of the basilar" syndrome. Neurology 30:72–79
5. Deleu D, Buisseret T, Ebinger G (1989) Vertical one-and-a-half syndrome: supranuclear downgaze paralysis with monocular elevation palsy. Arch Neurol 46:1361–1363
6. Ford CS, Schwartze GW, Weaver RG, Troost BT (1984) Monocular elevation paresis caused by an ipsilateral lesion. Neurology 34:1264–1267
7. Hommel M, Bogousslavsky J (1991) The spectrum of vertical gaze palsy following unilateral brainstem stroke. Neurology 41:1229–1234
8. Keane JR (1981) Sustained upgaze in coma. Ann Neurol 9:409–412
9. Keller EL, Heinen SJ (1991) Generation of smooth-pursuit eye movements: neuronal mechanisms and pathways. Neurosci Res 11:79–107
10. Kömpf D, Oppermann J (1986) Vertical gaze palsy and thalamic dementia. Neuro-ophthalmology 6:121–124
11. Kömpf D, Pasik K, Pasik P, Bender MB (1979) Downward gaze in monkeys. Stimulation and lesion studies. Brain 102:527–558
12. Kömpf D, Erbguth F, Kreiten K, Druschky KF, Hacke W (1987) Bilateral third nerve palsy in basilar vertebral artery disease. Neuroophthalmology 6:355–362
13. Leigh RJ, Zee DS (1991) The neurology of eye movements. Davis, Philadelphia
14. Pierrot-Deseilligny C, Rosa A, Masmoudi K, Rivaud S, Gaymard B (1991) Saccade deficits after a unilateral lesion affecting the superior colliculus. J Neurol Neurosurg 54:1106–1109
15. Susac JO, Hoyt WF, Daroff RB, Lawrence W (1970) Clinical spectrum of ocular bobbing. J Neurol Neurosurg Psychiatry 33:771–775
16. Thier P, Bachor A, Faiss J, Dichgans J, Koenig E (1991) Selective impairment of smooth-pursuit eye movements due to an ischemic lesion of the basal pons. Ann Neurol 29:443–448

The Internuclear Ophthalmoplegias

F.K. Thömke

Neurologische Klinik, Universität Mainz, Langenbeckstraße 1, 55131 Mainz, Germany

Introduction

Internuclear ophthalmoplegia (INO), which is caused by an ipsilateral medial longitudinal fasciculus (MLF) lesion, is characterized by adduction paresis of lateral gaze, usually with spared convergence [1–4]. In the opposite eye, abduction nystagmus and hypermetric abduction saccades are the main clinical and electro-oculographic abnormalities [1, 5, 6]. The origin of both is still debated. Abduction nystagmus has been explained by (a) an additional horizontal gaze paresis [7]; (b) vergence mechanisms aimed at alignment of the visual axes [8]; (c) interruption of descending excitatory projections from oculomotor nucleus internuclear neurons to contralateral abducens nucleus motoneurons [9]; (d) impaired inhibition of the medial rectus muscle of the abducting eye [10, 11]; or (e) adaptation to contralateral medial rectus undershoot [12]. Hypermetric abduction saccades reflect an increased phasic innervation [13, 14] thought to (a) account for the medial rectus paresis of the contralateral eye [12, 15] or (b) overcome an increased traction of the ipsilateral medial rectus due to deficient inhibition of the tonic resting activity during abduction movements [11]. Slowing of abduction saccades is another frequent additional feature in INO [11, 16–19] that has been attributed to various conditions such as (a) horizontal gaze paresis [16–18], (b) impaired medial rectus inhibition [11, 19], or (c) infranuclear intrapontine sixth nerve lesions [20]. The whole spectrum of ocular motor abnormalities associated with adduction paresis in INO, i.e., abduction nystagmus, hypermetric abduction saccades, and slowed abduction saccades, cannot be explained by any of the concepts mentioned. Previous approaches mostly focused on individual aspects, disregarding others, or were based on the findings in only a few patients.

Patients

The data reported are based on direct-current electro-oculography (EOG) of 150 consecutive patients with INO (78 unilateral, 72 bilateral). The

L.R. Caplan and H.C. Hopf (Eds.)
Brain-Stem Localization and Function

Table 1. Etiology of internuclear ophthalmoplegia in 150 consecutive patients (78 unilateral, 72 bilateral)

	Unilateral INO (n = 78)		Bilateral INO (n = 72)	
	n	%	n	%
Multiple sclerosis	38	48.7	49	68.0
Brain-stem infarction	37	47.4	21	29.2
Cancer	3	3.9	2	2.8

etiologies are shown in Table 1. Clinically, INOs were graded severe when adduction on lateral gaze was restricted, or mild when adduction saccades were slowed but unrestricted. Unilateral INO was mild in 41 patients and severe in 37. Bilateral INO was bilaterally mild in 13 patients, mild in one and severe in the other eye in 32, and bilaterally severe in 27.

The masseter and blink reflexes were recorded in 100 of the 150 patients. Trigeminal nerve involvement was excluded by clinical means (no masseter muscle paresis, no sensory disturbances of the three branches of the fifth nerve, no corneal reflex impairment). No patient had signs or symptoms of previous brain-stem involvement.

Electrophysiological Techniques

For EOG, horizontal eye movements were recorded separately in each eye with surface electrodes placed near the outer and inner canthi of the eyes. Bandwidth of the recording system was 0–40 Hz. All signals were documented by a linear ink jet writer. Voluntary horizontal saccades were obtained by following a light spot projected at random intervals on a screen 1.2 m in front of the patient. Peak velocities of 6 saccades of 30 deg to either side were evaluated graphically (paper speed: 100 mm/s). The normal range in 40 controls was 320–640 deg/s for 30 deg abduction saccades and 335–670 deg/s for 30 deg adduction saccades. Interocular differences did not exceed 35 deg/s.

The masseter reflex (MassR) was elicited by a brisk tap on the patient's jaw using a reflex hammer. The recording was triggered at the moment of the mechanical tap by a signal from a piezo-electric element mounted in the hammer. The MR was recorded using surface electrodes, with the recording electrode placed over the belly of the masseter muscle (25 mm above the margin of the mandible) and the reference over the jugular bone at the lateral edge of the orbit. Ten successive events were averaged.

The blink reflex (BR) was elicited by stimulating the supraorbital nerve on each side using rectangular stimuli of 0.1 ms duration, constant current of 25 mA, and intervals of 10 s. Responses from the orbicularis oculi muscles were recorded using surface electrodes, with the recording electrode placed

Table 2. Prevalence of abduction nystagmus and hypermetric abduction saccades in 78 consecutive patients with unilateral and 72 with bilateral internuclear ophthalmoplegia

	Unilateral INO				Bilateral INO			
	Mild (*n* = 41)		Severe (*n* = 37)		Mild (*n* = 58)		Severe (*n* = 86)	
	n	%	*n*	%	*n*	%	*n*	%
Abduction nystagmus with	19	46.3[a]	34	91.9	27	46.6[a]	82	95.3
hypermetric abduction saccades	12		32		23		81	
normometric abduction saccades	7		2		4		2	
Hypermetric abduction saccades	22	53.7[b]	35	94.6	38	65.5[a]	83	96.5
with abduction nystagmus	12		32		23		81	
without abduction nystagmus	10		3		15		2	

Values are calculated to the number of paretic eyes, i.e., 78 in 78 patients with unilateral and 144 in 72 patients with bilateral internuclear ophthalmoplegia.
[a] Significant ($p < 0.001$) difference between mild and severe INO.
[b] Significance $p < 0.01$.

inferior to the lower lid, halfway between the inner and outer edge of the orbit. Five consecutive events were evaluated. The ipsilateral and contralateral R_2 components were within normal limits in all our patients.

Electrophysiological Findings

Abduction Nystagmus and Hypermetric Abduction Saccades

Among patients with unilateral INO, 53 (67.9%) had abduction nystagmus and 57 (73.1%) had hypermetric abduction saccades in the eye contralateral to the MLF lesion. Their prevalences significantly increased with increasing severity of adduction paresis of the diseased eye (Table 2). In bilateral INO, abduction nystagmus was bilateral in 45 patients (62.5%) and unilateral in 15 (20.8%). Hypermetric abduction saccades were bilateral in 52 patients (72.2%) and unilateral in 13 (18.1%). Again, the prevalences significantly increased with increasing severity of adduction paresis (Table 2).

Slowing of Abduction Saccades

The prevalence of slowed abduction saccades in unilateral INO, their relation to the side of the MLF lesion, and their metricity are given in Table 3. Unilateral slowed abduction saccades were 95–255 deg/s slower than abduction saccades in the opposite eye (Fig. 1A,B). In patients with contralateral slowed abduction saccades, a horizontal gaze paresis was excluded by the high interocular velocity differences of adduction and abduction saccades

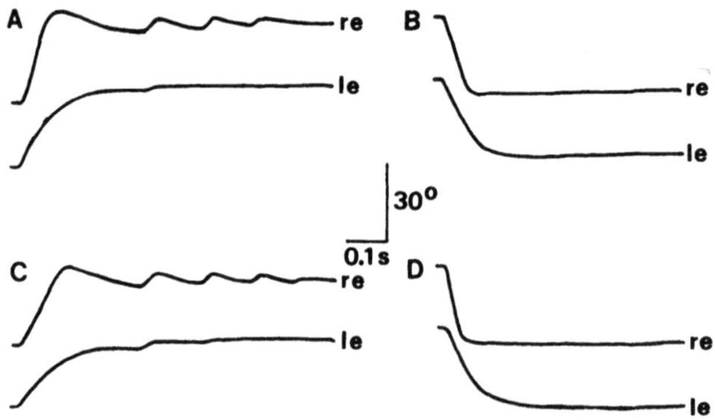

Fig. 1 A–D. Saccades from the primary position to the right (**A**) and left (**B**) in a patient with unilateral INO with slowed abduction saccades on the paretic eye. Saccades from the primary position to the right (**C**) and left (**D**) in a patient with unilateral INO with bilaterally slowed abduction saccades. *re*, right eye; *le*, left eye

Table 3. Prevalence of slowed abduction saccades in 78 consecutive patients with unilateral and 72 with bilateral internuclear ophthalmoplegia

	Unilateral INO $n = 78$			Bilateral INO $n = 72$		
	n		%	*n*		%
Unilaterally slowed abduction saccades	32		41.0	34		47.2
Ipsilateral to the MLF lesion and normometric		27				
Contralateral to the MLF lesion and hypermetric		4				
Contralateral to the MLF lesion and normometric		1				
Hypermetric					22	
Normometric					10	
Hypometric					2	
Bilaterally slowed abduction saccades	4		5.1	8		11.1
Ipsilateral to the MLF lesion normometric and contralateral to the MLF lesion hypermetric		4				
Bilaterally hypermetric					7	
Bilaterally normometric					1	

(155–200 deg/s). Bilateral slowed abduction saccades were 120–215 deg/s slower than adduction saccades in the eye contralateral to the MLF lesion (Fig. 1C,D). A one-and-a-half syndrome with conjugate saccadic slowing on gaze towards the side of the MLF lesion was observed in 11 patients (14.1%).

The prevalence of slowed abduction saccades in bilateral INO and their metricity are also given in Table 3. Bilateral slowed abduction saccades were 145–250 deg/s slower than abduction saccades in the opposite eye (Fig.

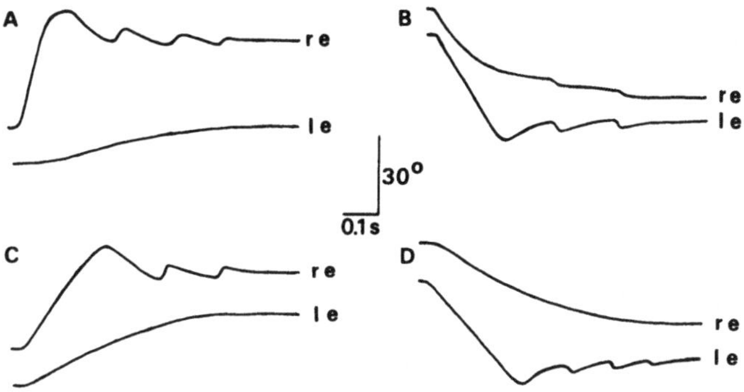

Fig. 2 A–D. Saccades from the primary position to the right (**A**) and left (**B**) in a patient with asymmetric bilateral INO with slowed abduction saccades on the more severely affected eye. Saccades from the primary position to the right (**C**) and left (**D**) in a patient with symmetric bilateral INO with bilaterally slowed abduction saccades. *re*, right eye; *le*, left eye

2A,B). Bilateral slowed abduction saccades with velocities between 135 and 305 deg/s were 100–180 deg/s faster than adduction saccades on the respective opposite eyes (Fig. 2C,D). A horizontal gaze paresis was excluded by the high interocular velocity differences between adduction and abduction saccades (100–180 deg/s).

Masseter and Blink Reflex Abnormalities

The prevalence of pathologic alterations of the MassR and/or the R_1 component of the BR (BR-R_1) in patients with unilateral and bilateral INO are given in Table 4. The clinical course was followed in 51 patients, 29 with unilateral and 22 with bilateral INO. On re-examination after 2–8 weeks, clinical recovery was always accompanied by improvement of the MassR and BR-R_1 responses, suggesting that INO and the reflex abnormalities were due to the same actual lesion.

INO and Clinical Diagnosis

Among our patients, 48.7% with unilateral INO had multiple sclerosis and 29.2% with bilateral INO had lacunar brain-stem infarction. These figures resemble those of Fötzsch [5] (68.3% multiple sclerosis in unilateral INO) and Gonyea [21] (29.3% brain-stem ischemia in bilateral INO). They contradict previous assumptions that unilateral INO usually originates from ischemic lesions [1, 22] and bilateral INO from multiple sclerosis [22].

Table 4. Prevalence of pathologic alterations of the masseter reflex and/or the R_1 component of the blink reflex (BR-R_1) in 60 patients with unilateral and 40 with bilateral internuclear ophthalmoplegia

	Unilateral INO $n = 60$			Bilateral INO $n = 40$	
	n		%	n	%
Unilateral abnormalities	46		76.7	15	37.5
Masseter reflex	31		51.7	14 (9)	35.0
Ipsilateral to the MLF lesion		28 (12)[a]			
Contralateral to the MLF lesion		3 (1)			
Blink reflex R_1	2		3.3		
Ipsilateral to the MLF lesion		2 (2)			
Contralateral to the MLF lesion					
Masseter reflex and blink reflex R_1	13		21.7	1	2.5
Ipsilateral to the MLF lesion		12 (7)			
Contralateral to the MLF lesion		1 (1)			
Bilateral abnormalities	7		11.7	20	50.0
Masseter reflex	7	(6)	11.7	10 (7)	25.0
Blink reflex R_1				2 (1)	5.0
Masseter reflex and blink reflex R_1				8 (5)	20.0

[a] Number of re-examined patients in parentheses.

Localization of the Lesion by Masseter and Blink Reflexes

The MassR is mediated by the trigeminal mesencephalic tract and nucleus (TMT & N) [23, 24]. MLF lesions extending slightly dorsolaterally approach the TMT & N and may involve the MassR arc [25]. The course of the trigeminofacial connection for the BR-R_1 is not exactly known. It probably runs from the trigeminal pontine access ventrolateral of the MLF to the facial nucleus, passing the principal sensory trigeminal nucleus close by [26, 27]. If peripheral causes of MassR or BR-R_1 impairment are excluded, as in all our patients, an abnormal MassR indicates an ipsilateral upper pontine or mesencephalic lesion [23, 24]. A BR-R_1 abnormality reveals an ipsilateral small lesion of the upper to mid third of the pons [26, 28].

The length of the MLF between the abducens and oculomotor nucleus is about 26 mm in man [25]. The principal sensory trigeminal nucleus (PSTN) extends over 6 mm. The space between the PSTN and the oculomotor nucleus is approximately 17 mm [25]. If lesions affecting the MLF are randomly scattered along its entire length, 23% of them are at the level of the PSTN and 65% rostral to it. Our figures are very close to these estimates. In 21.7% (unilateral INO) and 22.5% (bilateral INO) BR-R1 changes indicated a lesion at the PSTN level and in 63.3% (unilateral INO) and 60.0% (bilateral INO) isolated MassR abnormalities indicated a midbrain lesion. The functional significance of the MassR and BR-R_1 changes is illustrated by simultaneous electrophysiological and clinical recovery as observed in all 51 re-examined patients. The distribution derived from electrophysiological examinations and chance calculations corresponds to postmortem examinations which demonstrated lesions rostral to the abducens nucleus level in nine of ten patients [2–4, 21, 29–31]. In our experience, localization utilizing MassR and BR-R1 is even superior to magnetic resonance imaging (MRI). It was impossible to identify the causative lesion in INO, since MRI showed more than one lesion at different levels in 22 of 28 patients with unilateral and bilateral INO [18].

Slowed Abduction Saccades in INO

Slowed abduction saccades occurred in 46.2% and 58.3% of our patients with unilateral and bilateral INO, respectively. This agrees to previous observations [16–19]. A lesion of the infranuclear intrapontine sixth nerve, as proposed to cause slowed abduction saccades in INO [20], is not likely: (a) Autopsies [2–4, 21, 29–31] and brain-stem reflex studies in our patients demonstrated rostral pontine or midbrain lesions in almost 90% of those with unilateral or bilateral INO. (b) Simultaneous involvement of the MLF and infranuclear sixth nerve by a single lesion only occurs at the abducens nucleus level which, however, causes additional ipsiversive horizontal gaze paresis (one-and-a-half syndrome) [32, 33] seen in only 7.3% of our patients. (c) Excitation of the lateral rectus muscle is unimpaired in INO [9, 34–36].

There is also evidence against a horizontal gaze paresis [by additional lesions of the paramedian pontine reticular formation (PPRF) or the abducens nucleus] as the cause of slowed abduction saccades [16–18]: (a) Postmortem [2–4, 21, 29–31] and brain-stem reflex studies in our patients demonstrated lesions rostral to the PPRF and the abducens nucleus in the majority of unilateral or bilateral INO. (b) Slowed abduction saccades in unilateral INO occur predominantly in the paretic eye (84.4% of our patients) without accompanying slowing of adduction saccades in the contralateral eye. (c) Hypermetricity of saccades, as found in 69.0% of our patients with slowed abduction in bilateral INO, reflects increased phasic innervation [13, 14], generated by PPRF neurons and mediated via abducens nucleus motoneurons [37]. Damage to either the PPRF or the abducens nucleus, however, results in hypometric but not hypermetric slowed abduction saccades.

We think that impaired inhibition of the tonic resting activity of the antagonists, resulting in eye movement pareses despite normal excitation of the agonists [34, 38–40], is the most likely mechanism causing slowed abduction saccades in INO. Medial rectus inhibition is impaired in INO and always occurs in the paretic eye in unilateral INO [9, 34–36, 41]. Correspondingly, slowed abduction saccades occur predominantly in the paretic eye in unilateral and in the more severely affected eye in asymmetric bilateral INO (84.4% and 76.2% of our patients, respectively). Inhibitory MLF fibers crossing at the oculomotor nucleus level to contralateral medial rectus motoneurons, as proposed by Pola and Robinson [10], have not been demonstrated so far [42, 43]. Injury to such fibers would not impair inhibition of the medial rectus muscle ipsilateral to the MLF lesion [9, 35, 36].

There is no evidence that medial rectus inhibition is achieved by inhibition of abducens nucleus internuclear neurons mediated by inhibitory burst neurons of the contralateral reticular formation (disfacilitation) as supposed by Pierrot-Deseilligny [44] and Scudder et al. [45]: (a) Loss of excitatory activity of MLF fibers to medial rectus motoneurons in INO is not followed by decreased tonic resting activity of the medial rectus muscle [9, 34–36, 41]. (b) Experimental unilateral INO does not cause slowed abduction saccades in the paretic eye [46]. (c) Unilateral stimulation of inhibitory burst neurons evokes abduction of the ipsilateral eye only when the eye is adducted, but not in the primary or abducted positions, when there is no MLF fiber activity (Scudder 1989, personal communication). (d) Bilateral interruption of the MLF and the inhibitory projections to the abducens nuclei results in bilateral INO and bilateral loss of lateral rectus inhibition, but in only mild impairment of medial rectus inhibition [39].

However, there is some evidence for a separate homolateral inhibitory input to oculomotor nucleus neurons. Unilateral stimulation of the pontine reticular formation after bilateral MLF interruption evokes monosynaptic inhibitory potentials in ipsilateral oculomotor nucleus neurons [47] and ipsilateral medial rectus motoneurons [48]. Inhibitory neurons projecting to ipsilateral medial rectus motoneurons have been located through the entire n. reticularis pontis oralis and rostral n. reticularis pontis caudalis within the ipsilateral paramedian pontine tegmentum [48]. Fiber degeneration [49, 50] and autoradiographic studies [42, 51, 52] demonstrated an uncrossed connection between the pontine reticular formation and the oculomotor nucleus. These fibers ascend close to the midline, adjacent to but clearly separate from the MLF [42, 50], and come close to medial rectus motoneurons [52]. They originate from the same neurons within the ipsilateral paramedian pontine tegmentum that generate inhibitory monosynaptic potentials in ipsilateral medial rectus motoneurons [53]. This connection, previously called para-MLF [50], most likely mediates medial rectus inhibition (Fig. 3).

INO of Abduction Due to Impaired Medial Rectus Inhibition

Isolated lesions of the para-MLF may result in a prenuclear abduction paresis, which is due to impaired inhibition of the tonic resting activity of the ipsilateral medial rectus muscle. The appropriate name of such an eye movement abnormality is internuclear ophthalmoplegia of abduction (INO-abd) since the para-MLF is an internuclear connection [54]. This eye movement abnormality corresponds to the posterior internuclear opthalmoplegia

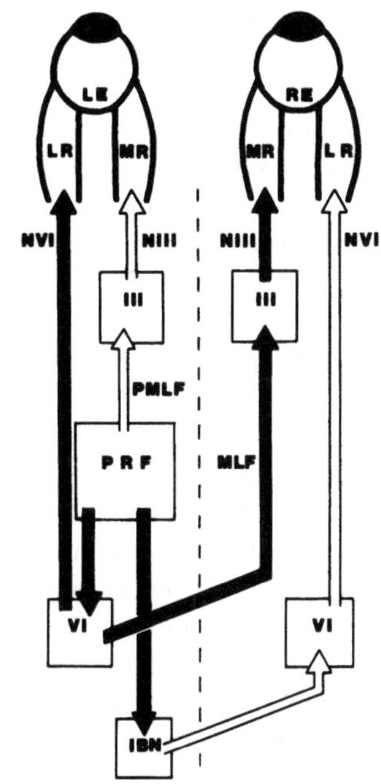

Fig. 3. Hypothetic horizontal gaze inhibitory system. The pattern of excitation and inhibition is shown for leftward gaze: *black arrows*, excitation; *white arrows*, inhibition; *LE*, left eye; *RE*, right eye; *LR*, lateral rectus; *MR*, medial rectus; *N VI*, abducens nerve; *VI*, abducens nucleus; *N III*, oculomotor nerve; *III*, oculomotor nucleus; *PRF*, pontine reticular formation; *MLF*, medial longitudinal fasciculus; *PMLF*, para-MLF; *IBN*, inhibitory burst neurons

of LUTZ. We considered INO-abd in nine patients with unilateral and six with asymmetric bilateral abduction paresis. In all of them, a rostral pontine or midbrain lesion was documented by isolated MassR abnormalities (ten patients) or combined MassR and BR-R_1 abnormalities (five patients). The lesion was ipsilateral to either the paretic or the more severely affected eye in patients with unilateral or asymmetric bilateral abduction paresis, respectively [54]. Improvement of the eye movement abnormality was always accompanied by improvement of the reflex changes, indicating that the same lesion was responsible for both abnormalities. Abduction paresis may occur with ipsilateral midbrain lesions, as was observed in mesencephalotomy [55] and mesodiencephalic hemorrhage [56]. Clinical signs associated with abduction paresis as upgaze paresis [57] and paralysis of vertical gaze and convergence [58, 59] also indicated the midbrain location of the lesion. The proposed pathophysiological mechanism was evident in one patient with unilateral abduction paresis [34]. He showed normal lateral rectus excitation but grossly impaired medial rectus inhibition [34]. Ipsilateral paresis of convergence indicated a midbrain lesion.

INO of Adduction Due to Impaired Lateral Rectus Inhibition

Following a deep midline cut at and below the abducens nucleus level, damage to the MLF was missed occasionally despite clinical signs of INO

[39]. The localization of the experimental lesion was likely to affect projections from inhibitory burst neurons to abducens nucleus motoneurons. According to our concept (Fig. 3), this would lead to loss of lateral rectus inhibition with normal medial rectus excitation as observed in these experiments [39]. Since a corresponding observation was reported in man [38] INO may occasionally occur with caudal pontine lesions interrupting inhibitory projections to lateral rectus motoneurons sparing the MLF.

Abduction Nystagmus in INO

Abduction nystagmus in INO has been attributed to (a) additional horizontal gaze paresis [7], (b) vergence mechanisms aimed at aligning the visual axes [8], (c) impaired excitatory projections from oculomotor nucleus internuclear neurons to contralateral abducens nucleus motoneurons [9], and (d) impaired medial rectus inhibition in the abducting eye [10]. Such mechanisms, without exceptions, would induce hypometric abduction saccades in the abducting eye. However, most patients with abduction nystagmus show hypermetric and, less frequently, normometric abduction saccades (Table 1). Other mechanisms that have been suggested are incomplete: An increased phasic innervation adjusted to impaired medial rectus inhibition of the abducting eye [11] is not a likely cause of unilateral INO, since medial rectus inhibition is impaired in the eye opposite to that showing abduction nystagmus [9, 34–36, 41]. Also, abduction nystagmus disappears after patching, for several days, the eye opposite to that exhibiting abduction nystagmus [15, 60, 61]. An increased phasic innervation adjusted to overcome the medial rectus paresis [12] will lead to accelerated hypermetric abduction saccades [13]. This mechanism cannot explain abduction nystagmus with (a) slowed hypermetric abduction saccades as observed in 35.4% of our patients with bilateral INO and (b) normometric abduction saccades as observed in 11.5% of our patients with unilateral and 6.9% with bilateral INO.

Thus, none of the above-mentioned concepts appropriately explains the spectrum of abduction nystagmus in INO. However, the combined effect of increased phasic innervation adjusted to the paretic eye and impaired medial rectus inhibition ipsilateral to the MLF lesion would do so.

According to Hering's law of equal innervation, an increased phasic innervation adjusted to the medial rectus paresis simultaneously activates the lateral rectus muscle of the opposite eye [12]. This results in hypermetric abduction saccades with nasal backdrift, which induces the generation of refixation saccades [12, 15]. This sequence as the cause of abduction nystagmus [12, 15] is confirmed by the occurrence of abduction nystagmus with hypermetric abduction saccades in the eye opposite to third nerve palsy [61] and contralateral to surgical weakening of the medial rectus muscle [60], as well as adduction nystagmus with hypermetric adduction saccades in the eye opposite to sixth nerve palsy [62]. Obviously, the increment of phasic innervation is adjusted to the degree of paresis of the involved muscles. Dissociated nystagmus with hypermetric saccades disappears after patching of the paretic eye for several days in all these conditions [15, 60–62]. We observed increasing prevalence of hypermetric abduction saccades with increasing severity of contralateral adduction paresis. A certain amount of increased phasic innervation seems to be a prerequisite to abduction nystagmus, since abduction nystagmus was missed despite hypermetric abduction saccades in some patients, preferentially with mild INO. This mechanism adequately explains abduction nystagmus with hypermetric abduction saccades, provided that abduction saccades are not slowed.

Superimposition of increased phasic innervation and medial rectus inhibition impairment best explains abduction nystagmus with slowed and hypermetric (9.0% of our patients with unilateral and 34.8% of those with bilateral INO) or normometric abduction saccades (11.5% of our patients with unilateral and 6.9% of those with bilateral INO) [11]. Metricity and velocity of abduction saccades under such conditions depend on the amount of increased phasic innervation in relation to the degree of impaired medial rectus inhibition. Such a mechanism requires associated lesions of the MLF and opposite para-MLF. Bilateral midbrain lesions were evident from bilateral MassR abnormalities in 11.7% of our patients with unilateral INO which may have extended to the para-MLF opposite to the MLF lesion. A para-MLF involvement can be expected preferentially in bilateral INO because of the close vicinity of both the MLF and the para-MLF.

References

1. Cogan D (1970) Internuclear ophthalmoplegia, typical and atypical. Arch Ophthalmol 84:583–589
2. Christoff N, Anderson PJ, Nathanson M, Bender MB (1960) Problems in anatomic analysis of lesions of the median longitudinal fascicuclus. Arch Neurol 2:293–304
3. Harrington RB, Hollenhorst RW, Sayre GP (1966) Unilateral internuclear ophthalmoplegia. Report of a case including pathology. Arch Neurol 15:29–34
4. Kupfer C, Cogan DG (1966) Unilateral internuclear ophthalmoplegia. A clinicopathological case report. Arch Ophthalmol 75:484–489
5. Fötzsch R (1971) Die internukleäre Ophthalmoplegie. Ophthalmologica 162:331–342
6. Crane TB, Yee RB, Baloh RW, Hepler RS (1983) Analysis of characteristic eye movement abnormalities in internuclear ophthalmoplegia. Arch Ophthalmol 101: 206–210
7. Bender MB, Weinstein EA (1939) Dissociated monocular nystagmus with paresis of horizontal ocular movements. Arch Ophthalmol 21:266–272
8. Stroud MH, Newman NM, Keltner JL, Gray AJ (1974) Abducting nystagmus in the medial longitudinal fasciculus (MLF) syndrome – internuclear ophthalmoplegia (INO). Arch Ophthalmol 92:2–5
9. Pierrot-Deseilligny C, Rigolet MH, Chain F (1979) Etude électromyographique de deux cas d'ophtalmoplégie internucléaire: déductions physiopathologiques. Rev Neurol (Paris) 135:143–152
10. Pola J, Robinson DA (1976) An explanation of eye movements seen in internuclear ophthalmoplegia. Arch Neurol 33:447–452

11. Feldon SE, Hoyt WF, Stark L (1980) Disordered inhibition in internuclear ophthalmoplegia. Analysis of eye movement recordings with computer simulations. Brain 103:113–137
12. Baloh RW, Yee RD, Honrubia V (1978) Internuclear ophthalmoplegia. I. Saccades and dissociated nystagmus. Arch Neurol 35:484–489
13. Bahill AT, Hsu FK, Stark L (1978) Glissadic overshoots are due to pulse width errors. Arch Neurol 35:138–142
14. Bahill AT, Troost BT (1979) Types of saccadic eye movements. Neurology 29: 1150–1152
15. Zee DS, Hain TC, Carl JR (1987) Abduction nystagmus in internuclear ophthalmoplegia. Ann Neurol 21:383–388
16. Bird AC, Leech J (1976) Internuclear ophthalmoplegia. An electro-oculographic study of peak angular saccadic velocities. Br J Ophthalmol 60:645–651
17. Caderas M, Meyer M (1981) Internucleäre Ophthalmoplegie und assoziierte oculomotorische Syndrome. Arch Psychiatr Nervenkr 230:17–30
18. Bronstein AM, Rudge P, Gresty MA, Du Boulay G, Morris J (1990) Abnormalities of horizontal gaze. Clinical, oculographic and magnetic resonance imaging findings. II. Gaze palsy and internuclear ophthalmoplegia. J Neurol Neurosurg Psychiatry 53:200–207
19. Thömke F, Hopf HC (1990) Abduction saccades in unilateral internuclear ophthalmoplegia. Neuroophthalmology 10:277–285
20. Henn V, Büttner U, Büttner-Ennever JA (1978) Supranukleäre Organisation der Okulomotorik – physiologische und anatomische Grundlagen. In: Kommerell G (ed) Augenbewegungsstörungen. Neurophysiologie und Klinik. Bergmann, Munich, pp 129–141
21. Gonyea EF (1974) Bilateral internuclear ophthalmoplegia. Association with occlusive cerebrovascular disease. Arch Neurol 31:168–173
22. Walsh FB, Hoyt WF (1969) Clinical neuro-ophthalmology, vol 2, 3rd edn. Williams and Wilkins, Baltimore, pp 239–247
23. Ongerboer de Visser BW (1983) Anatomical and functional organization of reflexes involving the trigeminal system in man. In: Desmedt JE (ed) Motor control mechanisms in health and disease. Raven, New York, pp 727–738
24. Hopf HC, Gutmann L (1990) Diabetic 3rd nerve palsy: evidence for a mesencephalic lesion. Neurology 40:1041–1045
25. Nieuwenhuys R, Voogd J, van Huijzen C (1988) The human central nervous system, 3rd edn. Springer, Berlin Heidelberg New York, pp 107–132
26. Kimura J (1970) Alteration of the orbicularis oculi reflex by pontine lesion. Arch Neurol 22:156–161
27. Ongerboer de Visser BW, Kuypers GJM (1978) Late blink reflex changes in lateral medullary lesions. An electrophysiological and neuroanatomical study of Wallenberg's syndrome. Brain 101:285–294
28. Kimura J (1983) Electrodiagnosis in diseases of nerve and muscle. Davis, Philadelphia, pp 323–351
29. Cogan DG, Kubik CH, Smith WC (1950) Unilateral internuclear ophthalmoplegia. Report on eight cases with one postmortem study. Arch Ophthalmol 44:783–796
30. Sahlgren E, Hofman-Bang E (1950) A case of internuclear ophthalmoplegia (Bielschowsky) with autopsy. Acta Psychiatr Neurol Scand 25:203–205
31. Ross AT, DeMyer WE (1966) Isolated syndrome of the medial longitudinal fasciculus in man. Anatomical confirmation. Arch Neurol 15:203–205
32. Meienberg O, Büttner-Ennever JA, Kraus-Ruppert R (1981) Unilateral paralysis of conjugate gaze due to lesion of the abducens nucleus. Neuroophthalmology 2:47–52
33. Pierrot-Desilligny C, Chain F, Serdaru M, Gray F, Lhermitte F (1981) The "one-and-a-half" syndrome. Electrooculographic analysis of five cases with deductions about the physiological mechanisms of lateral gaze. Brain 104:665–699
34. Orlowski WJ, Slomski P, Wojtowicz S (1965) Bielschowsky-Lutz-Cogan syndrome. Am J Ophthalmol 59:416–430

35. Loeffler JD, Hoyt WF, Slatt B (1966) Motor excitation and inhibition in internuclear palsy. Arch Neurol 15:664–671
36. Reuben RN, Gonzalez C (1964) Ocular electromyography in brain stem dysfunction. Clinical and electrophysiological analysis of a case with multiple eye movement disorders. Arch Neurol 11:265–272
37. Hepp K, Henn V, Vilis T, Cohen B (1989) Brainstem regions related to saccade generation. In: Wurtz R, Goldberg M (eds) The neurobiology of saccadic eye movements. Elsevier, Amsterdam, pp 105–212
38. Agnetti V, Traccis S, Depperu PV, Azzena GB (1981) Internuclear ophthalmoplegia of abduction: ocular EMG patterns. In: Huber A, Klein D (eds) Neurogenetics and neuro-ophthalmology. Elsevier, Amsterdam, pp 263–266
39. Burde RM, Lehman RAW, Roper-Hall G, Brooks J, Keltner JL (1977) Experimental internuclear ophthalmoplegia. Br J Ophthalmol 61:233–239
40. Pinhas I, Pinhas A, Goldhammer Y, Braham J (1978) Progressive supranuclear palsy: electromyographic examinations of eye muscles. Acta Neurol Scand 58:304–308
41. Gonzalez C, Reuben RN (1967) Ocular electromyography in the syndrome of the median longitudinal fasciculus. Patterns of inhibition and excitation. Am J Ophthalmol 64:916–926
42. Büttner-Ennever JA, Henn V (1976) An autoradiographic study of the pathways from the pontine reticular formation involved in horizontal eye movements. Brain Res 108:155–164
43. Büttner-Ennever JA, Akert A (1981) Medial rectus subgroups of the oculomotor nucleus and their abducens internuclear input in the monkey. J Comp Neurol 197:17–27
44. Pierrot-Deseilligny C (1985) Circuits oculomoteurs centraux. Rev Neurol (Paris) 141:349–370
45. Scudder CA, Fuchs AF, Langer TP (1988) Characteristics and functional identification of saccadic inhibitory burst neurons in the alert monkey. J Neurophysiol 59:1430–1454
46. Gamlin PDR, Gnadt JW, Mays LE (1989) Lidocaine-induced unilateral internuclear ophthalmoplegia: effect on convergence and conjugate eye movements. J Neurophysiol 62:82–95
47. Highstein SM, Cohen B, Matsunami K (1975) Monosynaptic projections from the pontine reticular formation to the IIIrd nucleus in the cat. Brain Res 75:340–344
48. Grantyn A, Grantyn R, Gaunitz U, Robiné KP (1980) Sources of direct excitatory and inhibitory inputs from the medial rhombencephalic tegmentum to lateral and medial rectus motoneurons in the cat. Exp Brain Res 39:49–61
49. Matano S (1970/1971) Experimental studies on the medial longitudinal fasciculus in the rabbit. V. Ascending fibers from the reticular formation and the oculomotor system. J Hirnforsch 12:241–253
50. Goebel HH, Komatsuzaki A, Bender MB, Cohen B (1971) Lesions of the pontine tegmentum and conjugate gaze paralysis. Arch Neurol 24:431–440
51. Graybiel AM (1977) Direct and indirect preoculomotor pathways of the brainstem: an autoradiographic study of the pontine reticular formation in the cat. J Comp Neurol 175:37–78
52. Büttner-Ennever JA, Miles TA, Henn V (1975) The role of the pontine reticular formation in oculomotor function. Exp Brain Res 23 Suppl:31
53. Remmel RS, Skinner RD, Pola J (1977) Cat pontomedullary reticular neurons projecting to the regions of the ascending MLF and the vestibular nuclei. In: Baker A, Berthoz A (eds) Control of gaze by brain stem neurons. Elsevier, Amsterdam, pp 163–166
54. Thömke F, Hopf HC, Krämer G (1992) Internuclear ophthalmoplegia of abduction. Clinical and electrophysiological data on the existence of an abduction paresis of prenuclear origin. J Neurol Neurosurg Psychiatry 55:105–111
55. Nashold BS, Gills JP (1967) Ocular signs from brain stimulation and lesions. Arch Opthalmol 77:609–618

56. Körney S (1959) Blickstörungen bei vasculären Herden des mesodiencephalen Übergangsgebietes. Arch Psychiatr Z Gesamte Neurol 198:535–543
57. Kommerell G (1975) Internuclear ophthalmoplegia of abduction. Isolated impairment of phasic ocular motor activity in supranuclear lesions. Arch Ophthalmol 93:531–534
58. Schiffter R (1975) Die internukleären Ophthalmoplegien. Klinische Analyse von 25 Kranheitsfällen. Nervenarzt 46:116–127
59. Weisberg LA (1986) Mesencephalic hemorrhages: clinical and computed tomographic correlations. Neurology 36:713–716
60. Von Noorden GK, Tredici TD, Ruttum M (1984) Pseudo-internuclear ophthalmoplegia after surgical paresis of the medial rectus muscle. Am J Ophthalmol 98:602–608
61. Abel A, Schmidt D, Dell'Osso LF, Daroff RB (1978) Saccadic system plasticity in humans. Ann Neurol 4:313–318
62. Kommerell G, Olivier D, Theopold H (1976) Adaptive programming of phasic and tonic components in saccadic eye movements. Invest Ophthalmol 15:657–660

Isolated Internuclear Ophthalmoplegia Following Head Injury: Magnetic Resonance Imaging and Neurophysiologic Findings

C. Strauss, W. Huk, and M. Emami

Abteilung für Neurochirurgie, Universität Erlangen-Nürnberg, Schwabachanlage 6, 91054 Erlangen, Germany

Introduction

Isolated unilateral internuclear ophthalmoplegia (INO) following head injury is rarely observed. A total of five patients presenting with INO as the only sign of brain-stem trauma have been reported in the literature [1, 4, 6, 7, 9].

INO is usually associated with cerebrovascular disorders or results from demyelinating disease. Rarely, it is observed in tumors or malformations such as the Arnold-Chiari syndrome [5]. In most circumstances INO is associated with other brain-stem signs [3, 5]. Due to this association, even magnetic resonance imaging (MRI) can fail to identify the exact topography of INO [5, 7]. This has also been observed when neurophysiological methods were used [8].

The case of isolated post-traumatic INO presented here illustrates the importance of clinical diagnosis, MRI, and neurophysiological methods in order to achieve acurate topograhical identification of the lesion actually responsible for the symptomatology.

Case Report

A 52-year-old woman was admitted following minor closed head injury. She complained of headache and double vision when looking to the left side. On admission the patient was alert and oriented but suffered from retrograde and anterograde amnesia.

She also complained of slight hearing loss on the left side. The physical examination revealed a hematotympanum on the left. On neurological examination a complete adduction deficit of the right eye was demonstrated, with associated horizontal nystagmus of the left eye on abduction. Convergence was not impaired. Otherwise, the neurological exam disclosed no further abnormalities.

L.R. Caplan and H.C. Hopf (Eds.)
Brain-Stem Localization and Function

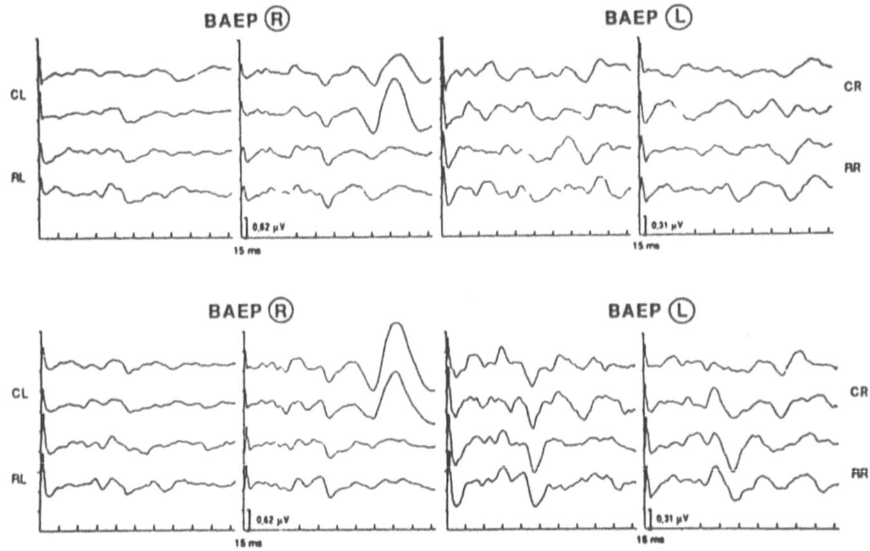

Fig. 1. Brain-stem auditory evoked potentials on admission (*above*) and 8 months after trauma (*below*). Recovery of amplitude reduction and reappearance of wave III after left-side stimulation on follow-up

Diagnostic Imaging Procedures

Plain skull X-rays showed a left parietal skull fracture reaching the mastoid bone. Computer tomography (CT) demonstrated a left otobasal fracture with fluid accumulation within the mastoid air cells. In addition, a subarachnoidal hemorrhage with blood accumulation on the tentorial edges was seen. On retrospective analysis, a small pontomesencephalic lesion was identified, possibly representing a hemorrhage. On magnetic resonance imaging (MRI) (1.5 Tesla) this lesion was verified (Fig. 2). It was localized at the level of the inferior colliculi ventrally and paramedian to the sylvian aqueduct on the right side. It was 2 mm in diameter. Based on hypointensity in proton and T_2-weighted images the diagnosis of a small brain-stem hemorrhage was made.

Neurophysiological Investigations

A pure-tone audiogram documented a pancochlear hearing loss of 50 dB on the left side, consistent with middle ear pathology. On vestibular testing no abnormalities were identified. Brain-stem auditory evoked potentials (BAEP) revealed a 50% amplitude reduction V-V' on stimulation of the left side (condensation 0.15 μV left – 0.51 μV right, rarefaction 0.20 μV left –

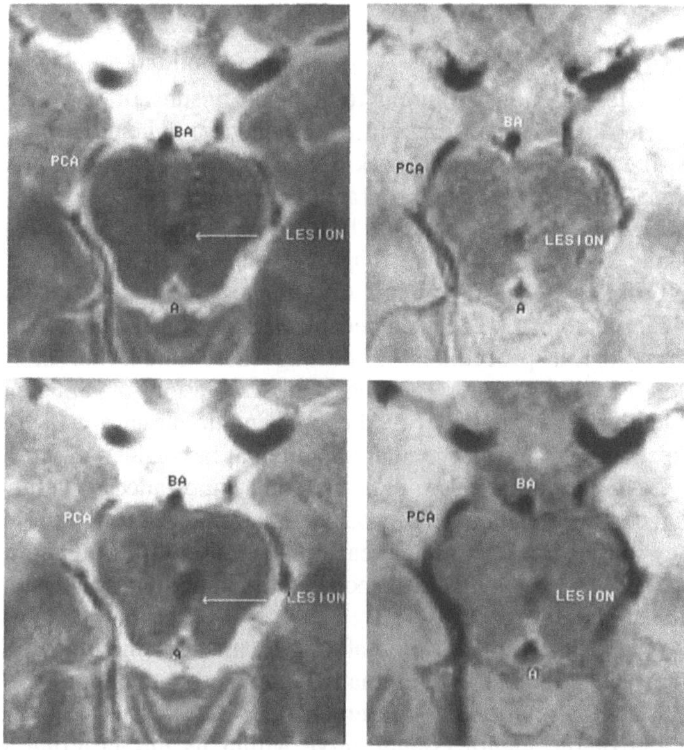

Fig. 2. Magnetic resonance imaging 2 weeks (*above*) and 14 months (*below*) after trauma, showing a small lesion at the level of inferior colliculi. Discrete enlargement of lesion on follow-up

0.41 µV right). Furthermore, wave III was unobtainable, and I–V interpeak latency was prolonged (0.66 ms condensation, 0.96 ms rarefaction). There were no abnormalities on right-sided stimulation (Fig. 1).

Follow-up

Hearing loss recovered completely. Four weeks after the trauma occurred, the adduction deficit had slightly improved to the extent of crossing the midline. The contralateral nystagmus remained unchanged. Symptoms persisted and remained practically unchanged at 14-month follow-up.

MRI showed no major changes; the lesion diameter was slightly increased to 3 mm on an axial view (Fig. 2).

BAEP recovered within 8 months. Reappearance of wave III after 2 months documented the initial I–V interpeak latency shift on the left side as being caused by a I–III interpeak latency shift.

Discussion

Internuclear ophthalmoplegia is rarely caused by trauma [1, 4, 6, 7, 9]. Such small lesions are attributed to ischemic insults or rupture of perforating arteries due to shear forces [6, 7]. Brain-stem ischemia has been emphasized as the etiology of stiff neck syndrome when symptoms were observed immediately following chiropractic maneuvers [3].

MRI in the case presented clearly documented a small pontomesencephalic paramedian hemorrhage as the probable cause of the isolated INO following head trauma. Additional clinical and subclinical brain-stem symptomatology were excluded by clinical examination and various neurophysiological procedures. Following these additional investigations, the isolated clinical and neurophysiological lesion of unilateral INO could roughly be correlated with the small pontomesencephalic lesion on MRI, representing the right medial longitudinal fasciculus.

BAEP findings also illustrate the importance of employing various diagnostic procedures. Pancochlear hearing loss on pure-tone audiogram was caused by a left middle ear effusion discovered on clinical examination and CT and can be held responsible for amplitude deterioration after left-side stimulation.

Reappearance of wave III after 2 months identified the initially prolonged interpeak latency I–V as actually being a prolongation of the I–III interval, rather than of the III–V interval. The increased I–III latency points to a lesion of the eighth cranial nerve within its intracisternal segment rather then an intrinsic brain-stem lesion. This is underlined by CT, showing subarachnoidal hemorrhage along the tentorial edges; this is typical for external force, which may be held responsible for stretching of nervous structures within the cerebello-pontine angle.

The case presented illustrates the necessity for a diagnostic battery that includes clinical examination, neurophysiological testing, and imaging techniques in order to correlate clinical findings with small brain-stem lesions visible on MRI and neuroanatomical structures.

References

1. Alexander JA, Castillo M, Hoffman JC jr (1991) Magnetic resonance findings in a patient with internuclear ophthalmoplegia. J Clin Neuroophthalmol 11(1):58–61
2. Atlas SW, Grossmann RI, Savino PJ, Schatz NJ, Sergott RC, Bosley TM, Hackney DB, Goldberg HI, Bilaniuk LT, Zimmerman RA (1987) Internuclear ophthalmoplegia: MR – anatomic relation. AJNR 8(2):243–247
3. Baker RS (1979) Internuclear ophthalmoplegia following head injury. J Neurosurg 51:552–555
4. Beck RW, Meckler RJ (1981) Internuclear ophthalmoplegia after head injury. Ann Ophthalmol 13(6):671–675
5. Bronstein AM, Rudge P, Gresty MA, Du Boulay G, Morris J (1990) Abnormalities of horizontal gaze. Clinical oculographic and magnetic resonance imaging findings.

II. Gaze palsy and internuclear ophthalmoplegia. J Neurol Neurosurg Psychiatry 53(3):200–207

6. Catalano RA, Sax RD, Krohel GB (1986) Unilateral internuclear ophthalmoplegia after head trauma. Am J Ophthalmol 101(4):491–493
7. Haller KA, Miller-Meeks M, Kardon R (1990) Early magnetic resonance imaging in acute traumatic internuclear ophthalmoplegia. Ophthalmology 97(9):1162–1165
8. Hopf HC, Thomke F, Gutmann L (1991) Midbrain vs pontine medial longitudinal fasciculus lesions: the utilisation of masseter and blink reflexes. Muscle Nerve 14(4): 326–330
9. Tanaka R, Kawakami K, Yoshioka N (1973) A case of syndrome of the medial longitudinal fasciculus following head injury. Clin Neurol 13:518–520

Nuclear and Fascicular Oculomotor Nerve Lesions in Brain-Stem Infarcts: A Clinicomorphological Study

W. Grisold, K. Jellinger, M. Drlicek, and D. Volc

Neurologische Abteilung, Kaiser Franz Josef-Spital, Ludwig Boltzmann-Institut, Baumgartnerhöhe 1, 1140 Wien, Austria

Introduction

Lesions of the oculomotor nerve nucleus are rare. Anatomically, the nuclear complex of the third nerve extends rostrocaudally, ventral to the aqueduct in the long axis of the midbrain. Dorsal and rostral to the main oculomotor nuclei is the unpaired autonomic Westphal-Edinger nucleus, and at the caudal level is the caudal central nucleus of the levator palpebrae superioris. The main part of the oculomotor complex consists of two paired columns that are topographically divided into dorsolateral, intermediate, and ventrolateral nuclei. The most medial subnuclei innervate the superior rectus muscle. Each medial subnucleus has fibers extending to the opposite eye, but decussation takes place within the nuclear complex. The decussating fibers for the superior rectus muscle traverse the contralateral subnucleus. However, the lateral oculomotor subnuclei have axons that extend to the ipsilateral eye fibers (dorsal subnucleus: inferior rectus muscle; intermediate subnucleus: inferior oblique muscle) Neurons supporting the medial rectus muscle are distributed in three different subnuclei [9].

Rostral to the oculomotor nerve complex are the structures for the vertical gaze (nucleus of the medial longitudinal fasciculus riMLF, posterior commissure, interstitial nucleus of Cajal, and nucleus of Darkewitsch [20, 25]. Possibly, the efferent riMLF pathway for up- and downgaze lies separately [32]; thus upgaze, downgaze, combined up- and downgaze, mononuclear elevation palsy, and the "vertical one-and-a-half syndrome" [5] may occur in association with rostral oculomotor nerve lesions.

The fascicular portion of the oculomotor nerve traverses the midbrain and is located next to the MLF, the superior cerebellar peduncle, and the red nucleus. Lesions of corticofugal and colliculofugal pathways which control horizontal saccades may cause additional palsy of contralateral horizontal saccades [29], due to damage to the smooth ocular pursuit pathways [22, 35]. Table 1 presents the features of oculomotor nuclear and fascicular lesions.

L.R. Caplan and H.C. Hopf (Eds.)
Brain-Stem Localization and Function

Table 1. Differential diagnostic considerations in oculomotor nerve nuclear and fascicular lesions (references in text)

1. *Nuclear lesion:*
   a) Unilateral oculomotor lesion with contralateral superior rectus muscle paresis and bilateral ptosis ("classical description")
   b) Bilateral oculomotor nerve lesion with or without INO and sparing of levator function
   c) Bilateral ptosis and sparing of rostral oculomotor nucleus
   d) Singular oculomotor nerve innervated muscles, except:
   Rectus superior muscle (crossed innervation)
   Levator palpebrae superioris muscle (single caudal subnucleus)
   Pupillary constrictors (various subnuclei dispersed)
   Medial rectus muscle (3 subnuclei)
2. *Fascicular lesion:*
   a) Complete oculomotor lesion – all contralateral muscles spared
   b) Isolated inferior oblique muscle paresis
   c) Unilateral dilated, fixed pupil
   d) Paresis of inferior oblique, superior rectus, medial rectus, levator palpebrae with sparing of inferior rectus muscle and pupil
   e) Paresis of inferior oblique, superior rectus, medial rectus, levator, inferior rectus muscles and pupillary sparing
3. *Fascicular lesions often accompanying nuclear lesions (paramedian arterial branches supply both structures):*
   Lethargy, bilateral ptosis, and unilateral oculomotor nerve lesion (suggestive of combined nuclear and fascicular lesion)
4. *Additional damage to neighboring structures causing complex syndromes:*
   a) Red nucleus (Claude syndrome)
   b) Subthalamic nucleus (Benedikt syndrome)
   c) Brachium conjunctivum below decussation: (Nothnagel syndrome)
   d) Peduncular lesion (Weber syndrome)
   e) Lesion of fibers at mesencephalic tegmentum: ipsilateral oculomotor nerve lesion III plus loss of control of horizontal saccades (restriction of abduction may, however, be overcome by oculocephalic reflex)
   f) Lesion of posterior comissure: upgaze palsy
   g) Lesion of subthalamic structures (riMLF, periaqueductal gray, nucleus Darkewich, nucleus Cajal) up- and down-gaze palsy
   h) Lesions of the territory in posterior thalamo- subthalamic paramedian artery supply, bilateral riMLF, periaqueductal gray: downgaze palsy
   i) Monocular elevation palsy: (superior rectus and inferior oblique muscle) ipsi- or contralateral
   j) Vertical "one-and-a-half" bilateral upgaze and unilateral downgaze palsy bilateral downgaze and monocular upgaze palsy
   k) Excessive convergence tone: "contralateral pseudo-abducens palsy"

Case Reports

Case 1. A 50-year-old man presented with intermittent attacks of dizziness and vertigo. Initially he had bilateral ptosis. His left pupil was dilated and fixed to light. The eyes were slightly divergent, more on the right than on the left. No voluntary eye movements were possible.

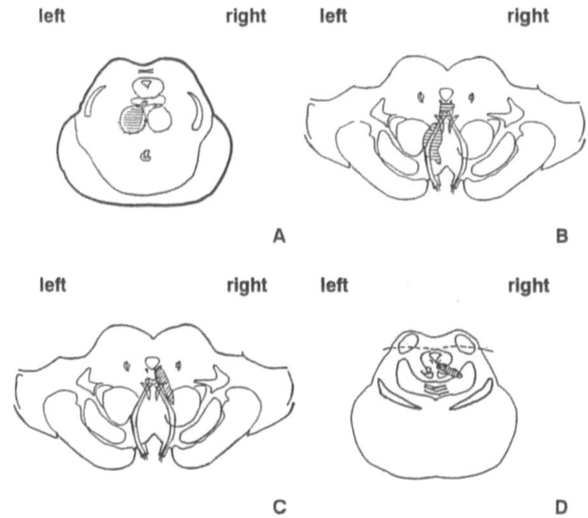

Fig. 1A–D. Location of lesions at different levels of the oculomotor nerve nucleus. **A** case 1; **B** case 2; **C** and **D** case 3

The patient's consciousness was mildly impaired at the onset but he was always able to answer questions adequately. He had difficulties with swallowing and had bilateral ataxia.

After 2 weeks, ptosis remained in the left eye, the left pupil was wider than the right, showing no reaction to light, and there was vertical up- and downgaze palsy. The convergence reaction could not be elicited. Left-side ptosis persisted for 1 year, as did dysarthria and ataxia. The patient survived for 4 years.

At autopsy the infracollicular level of the mesencephalon was examined (Fig. 1A). Here, the paramedian lesion extended from the caudal portion of the oculomotor nerve to the brachium conjunctivum.

Case 2. A 69-year-old hypertensive woman experienced an acute onset of mild impairment of consciousness, small non-reactive pupils and roving eye movements. Bilateral incomplete ptosis was noted, more accentuated on the left. There was only slight abduction of the left eye; the movement of the right eye was not impaired in rostral and caudal planes. MRI showed a small periaqueductal medial lesion in the midbrain tegmentum. The patient died 5 months later.

The plane of the superior colliculi was examined (Fig. 1B) neuropathologically. In addition to the oculomotor nuclear lesion, there was a fingerlike paramedian extension of the infarction ventrally into the oculomotor nerve fascicle and the medial part of the red nucleus. Myelinated nerve fiber regeneration was present in the proximal part of the oculomotor nerve fascicle, next to the nucleus (Fig. 2)

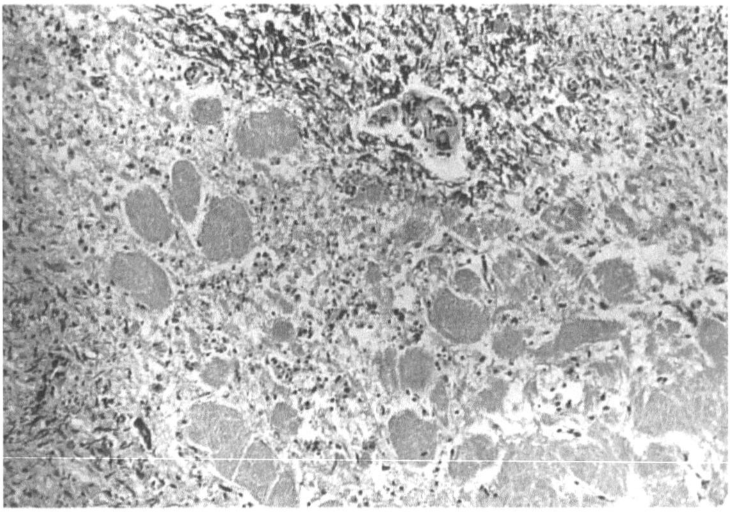

Fig. 2. Case 2: Myelinated nerve fiber regeneration in the proximal part of the oculomotor nerve fascicle next to the nucleus. Klüver-Barrera stain

Case 3. A 61-year-old man was admitted with dysarthria and swallowing difficulties. After a brief remission of neurological signs a mesencephalic stroke occurred. Findings were a complete oculomotor nerve lesion on the left, vertical up- and downgaze palsy, bilateral internuclear ophthalmoplegia, rotatory nystagmus and right-sided cerebellar ataxia. He survived one week.

Neuropathological examination (Fig. 1C and D) revealed a paramedian nuclear nerve lesion almost restricted to the the oculomotor nerve nucleus. Involvement of the MLF and sparing of the nerve fascicles and adjacent structures were noted. Unfortunately, the rostral extension of the lesion was not examined. There was an additional ipsilateral thalamopeduncular infarction.

Discussion

The cases reported demonstrate that nuclear oculomotor nerve lesions do not present a uniform clinical picture. Although two of our patients (cases 1 and 2) presented with ophthalmoplegia and bilateral ptosis, the involvement of pupillary fibers and of the brachium conjunctivum was responsible for different clinical features due to different morphological patterns of the paramedian lesions, although both the oculomotor nucleus and the nerve fascicle were involved. In case 2, additional nerve fiber regeneration of the fascicular part of the oculomotor nerve was observed despite a comparatively short (5 months) survival, while the patient in case 1, who survived for 5 years, did not show any signs of axonal regeneration.

Patient 3, despite a histological lesion limited to the oculomotor nucleus without involvement of the nerve fascicle, presented clinically with up- and downgaze palsy in addition to a complete unilateral oculomotor nerve lesion, while bilateral ptosis was absent, suggesting that the caudal part of the oculomotor nucleus was spared. Additional infarction of the ipsilateral thalamus suggests that subthalamic mesencephalic structures responsible for vertical gaze generation [20] were also affected.

Although not described as a "classical" brain-stem syndrome [13, 16, 28], the combination of impaired consciousness, bilateral ptosis, unilateral third nerve palsy, and contralateral paresis of the superior rectus muscle suggests an oculomotor nerve nuclear infarction [3, 11, 27]. The same applies for a bilateral third nerve palsy (with or without INO) and spared levator function [17], bilateral ptosis with sparing of other oculomotor functions [10, 19, 23], isolated superior rectus muscle paresis contralateral to the lesion [15, 34]. However, any extraocular eye muscle innervated by the oculomotor nerve may be affected in nuclear lesions, with the exception of the levator superioris muscle (fiber crossing within the nuclear complex), levator palpebrae muscles (a single subnucleus innervates both levator muscles), rectus medialis muscle (three subnuclei), and the pupillary constrictors (subnuclei are dispersed through the nuclear complex). Little information is available about involvement of structures concerned with the convergence reaction in nuclear nerve lesions [26]. Due to the rostral adjacent structures controlling vertical gaze [20], rostral oculomotor nerve lesions may be associated with vertical gaze palsies, as demonstrated in case 3. The possibility of a so-called prenuclear oculomotor nerve lesion must also be considered in this region [4, 6].

Clinically less relevant but noteworthy are eye movement abnormalities caused by lesions of neighboring ocular motor pathways [22, 29, 35]. Despite the anatomical and functional peculiarities of oculomotor nerve lesions, other causes of oculomotor nerve lesions have to be considered when evaluating a patient [2, 24, 34]. In conjunction with nuclear lesions, the importance of additional fascicular damage [7] must be emphasized as demonstrated by cases 1 and 2. This can be explained by the vicinity of the nuclear and fascicular structures and also by their vascular supply, as paramedian branches near the top of the basilary artery feed both structures [14]. Clinically, a complete oculomotor nerve lesion with sparing of the contralateral eye, damage to neighboring structures (red nucleus: Claude syndrome; subthalamic structures: Benedikt syndrome; brachium conjunctivum: Nothnagel syndrome; cerebral peduncle: Weber's syndrome), association with gaze abnormalities, convergence disorders, pseudoabducens paresis [30], and other rare findings [7] may occur. Isolated fascicular lesions of the oculomotor nerve are rare [23] but are characterized by its intraparenchymatous topographic arrangement [1, 12]. Thus an isolated paresis of the inferior oblique muscle [12], a unilateral dilated fixed pupil [33], paresis of the inferior oblique, superior rectus, medial rectus levator muscle with sparing of the inferior rectus muscle and pupil [31], and paresis of inferior oblique, superior rectus, medial rectus, levator, inferior rectus, and pupillary sparing [8] are suggestive of an isolated fascicular oculomotor nerve lesion.

Apart from structural concepts of the oculomotor nucleus, and effects of lesions on neigboring structures, especially the fascicular part, the time factor of lesions must also be considered. Rapid development of the neuro-ophthalmological deficits suggests vascular lesions, which, as demonstrated in case 3, may have sequelae. Progression often occurs in patients with basilar artery thrombosis; isolated INO has been observed in several cases as the presenting syndrome [18, 21].

References

1. Abdollah A, Francis GS (1990) Intraaxial divisional oculomotor nerve paresis suggests intraaxial fascicular organization (letter to the editor). Ann Neurol 28:589–590
2. Berlit P (1991) Isolated and combined pareses of cranial nerves III, IV and VI. J Neurol Sci 103:10–15
3. Biller J, Shapiro R, Evans LS, Haag JR, Fine M (1984) Oculomotor nuclear complex infarction. Arch Neurol 41:985–987
4. Bogousslavsky J, Meienberg O (1987) Eye movement disorders in brain stem and cerebellar stroke. Arch Neurol 44:141–148
5. Bogousslavsky J, Regli F (1984) Upgaze palsy and monocular paresis of downward gaze from ipsilateral thalamo-mesencephalic infarction: a vertical "one-and-a-half" syndrome. J Neurol 231:43–45
6. Bogousslavsky J, Regli FG, Ghika J (1983) Internuclear ophthalmoplegia, praenuclear paresis of contralateral superior rectus and bilateral ptosis. J Neurol 230:197–203
7. Brazis PW (1991) Localisation of lesions of the oculomotor nerve: recent concepts. Mayo Clin Proc 66:1029–1035
8. Breen LA, Hopf HC, Farris BK, Gutmann L (1991) Pupil-sparing oculomotor nerve palsy due to midbrain infarction. Arch Neurol 48:105–106
9. Büttner-Ennever JA, Akert K (1981) Medial subgroups of the oculomotor nucleus and their abducens internuclear input in the monkey. J Comp Neurol 197:17–21
10. Caplan LR (1974) Ptosis. J Neurol Neurosurg Psychiatry 37:1–7
11. Caplan LR (1980) "Top of the basilar" syndrome. Neurology (Nu) 30:72–79
12. Castro O, Johnson LN, Mamourian AC (1990) Isolated inferior oblique paresis from brain stem infarction. Arch Neurol 47:235–237
13. Currier RD (1975) Syndromes of the medulla oblongata. In: Vinken PJ, Bruyn GW (eds) Handbook of clinical neurology, vol 2. North-Holland, Amsterdam, pp 217–237
14. Duvernoy HM (1978) Human brainstem vessels. Springer, Berlin Heidelberg New York
15. Eustace P (1985) Partial nuclear third nerve palsies. Neuroophtalmology 5:259–262
16. Fog M, Hein-Sorensen O (1975) Mesencephalic syndromes. In: Vinken PJ, Bruyn GW (eds) Handbook of clinical neurology, vol 2. North-Holland, Amsterdam, pp 272–285
17. Glaser JS (1991) Neuroophtalmology, 2nd edn. Lippincott, Philadelphia
18. Grisold W, Drlicek M, Jellinger K, Volc D (1991) Ophthalmoneurologische Symptome bei Basilaristhrombosen. Intensive Notfallbehandlung 16:195–200
19. Growdon JH, Winkler GF, Wray SH (1974) Midbrain ptosis.Arch Neurol 30:179–181
20. Hommel M, Bogousslavsky J (1991) The spectrum of vertical gaze palsy following unilateral brain-stem stroke. Neurology 41:1229–1234
21. Jellinger K (1986) Neuropathology and clinical signs of brainstem disorders. In: Kunze K, Zangemeister WH, Arlt A (eds) Clinical problems of brainstem disorders. Thieme, Stuttgart, pp 17–29

22. Kato I, Watanabe J, Nakamura T, Harada K, Hasegawa T, Kanayama R (1990) Mapping of brainstem lesions by the combined use of tests of visually induced eye movements. Brain 113:921–935
23. Keane JR (1988) Isolated brain-stem third nerve palsy. Arch Neurol 45:813–814
24. Keane JR, Zaias B, Itabashi HH (1984) Levator-sparing oculomotor nerve palsy caused by a solitary midbrain metastasis. Arch Neurol 41:210–212
25. Leigh RJ, Zee DS (1983) The neurology of eye movements. Davies, Philadelphia
26. Lindner K, Grisold W (1992) Transient dissociated unilateral convergence paralysis in a patient with thalamo-tectal haemorrhage (case report). J Neurol Neurosurg Psychiatry 55:731–733
27. Liu GT, Carrazana EJ, Charness ME (1991) Unilateral oculomotor palsy and bilateral ptosis from paramedian midbrain infarction. Arch Neurol 48:983–986
28. Loeb C, Meyer JS (1975) Pontine syndromes. In: Vinken PJ, Bruyn GW (eds) Handbook of clinical neurology vol 2. North-Holland, Amsterdam, pp 238–271
29. Masdeu JC, Rosenberg M (1987) Midbrain-diencephalic horizontal gaze paresis. J Clin Neuroophthalmol 7:227–234
30. Masdeu J, Brannegan R, Rosenberg M, Dobben G (1980) Pseudo abducens palsy with midbrain lesion (Abstr). Ann Neurol 8:103.
31. Nadeau SE, Trobe JD (1983) Pupil sparing in oculomotor palsy: a brief review. Ann Neurol 13:143–148
32. Pierrot-Deseilligny C (1985) Curcuits oculomoteurs centraux. Rev Neurol (Paris 141:349–370
33. Shuaib A, Israelian G, Lee MA (1989) Mesencephalic hemorrgage and unilateral pupillary defect. J Clin Neuroophthalmol 9:47–49
34. Warren W, Burde RM, Klingele RT, Roper-Hall G (1982) Atypical oculomotor paresis. J Clin Neuroophtalmol 2:13–18
35. Zackon DH, Sharpe JA (1984) Midbrain paresis of horizontal gaze. Ann Neurol 16:495–504

A Case Report Illustrating the Brain-Stem Anatomy of Horizontal Eye Movements

K. Tiel-Wilck, T. Lempert, and J. Schultes

Abteilung für Neurologie, Universitätsklinikum Rudolf Virchow,
Freie Universität Berlin, Spandauer Damm 130, 14050 Berlin, Germany

Case Report

Stage 1: Horizontal Gaze Palsy

A 31-year-old man developed postural imbalance and 3 days later a horizontal gaze palsy to the left. He was unable to perform saccadic, pursuit, or vestibular eye movements beyond the midline to the left. Saccades from the right back to the midposition were slightly slowed. There was a gaze-evoked nystagmus to the right. Simultaneously with the oculomotor abnormalities, a left-sided infranuclear facial palsy with preserved taste and lacrimal secretion appeared.

Stage 2: One-and-a-half Syndrome

Three days later, the patient had oscillopsia on right gaze. An adduction weakness of the left eye with concomitant abduction nystagmus of the right eye was observed. Convergence was normal. A one-and-a-half syndrome was diagnosed, i.e., horizontal gaze palsy and ipsilateral internuclear ophthalmoplegia.

Stage 3: One-and-a-half Syndrome and Abducens Nerve Palsy

Five days later, the internuclear ophthalmoplegia persisted. The horizontal gaze palsy recovered: right eye adduction reached 20°. The left eye, however, was abducted only slightly beyond the midline to the left, indicating an additional abducens nerve palsy.

L.R. Caplan and H.C. Hopf (Eds.)
Brain-Stem Localization and Function

Brain-stem Anatomy of the Horizontal Oculomotor System

Abducens Nucleus and Its Efferent Pathways

The nucleus of the abducens nerve is the final common pathway for all types of conjugate horizontal eye movements. It contains two different populations of neurons, namely the abducens motoneurons, which innervate the lateral rectus muscle, and the abducens interneurons, which project to the contralateral oculomotor nucleus [2]. These interneurons cross the midline at the level of the abducens nucleus and travel upward within the medial longitudinal fasciculus (MLF), finally reaching the medial rectus subnucleus [9].

Afferent Pathways

The abducens nucleus receives its afferent input from supranuclear structures which generate the different types of eye movements.

Saccades. Saccadic burst neurons are located within the ipsilateral nucleus pontis centralis caudalis, which is a part of the paramedian pontine reticular formation (PPRF). Inhibitory saccadic neurons which project to the contralateral abducens nucleus lie adjacent and caudal to the abducens nucleus and the PPRF within the medullary reticular formation (MedRF). Simultaneous activation of these two types of neurons excites the ipsilaterally pulling eye muscles with concomitant inhibition of their antagonists [13, 14].

Vestibular Eye Movements. Stimulation of one horizontal semicircular canal activates neurons within the ipsilateral medial vestibular nucleus (MVN). These project to the contralateral abducens nucleus and induce compensatory eye movements which stabilize gaze in space. Inhibitory neurons of the MVN relate to the ipsilateral abducens nucleus, thus relaxing the antagonists [10].

Smooth-Pursuit Eye Movements. Two different pontine structures are involved in the generation of pursuit eye movements: the dorsolateral pontine nuclei (DLPN) receive cortical commands and project to the ipsilateral cerebellum; the MVN receives its input from the ipsilateral cerebellum. The efferent fibers of the MVN approach the ipsilateral abducens nucleus travelling in proximity to the caudal PPRF [6].

Gaze Holding. The tonic eye muscle innervation which is required for the maintenance of eccentric eye position is generated by the nucleus prepositus hypoglossi (NPH), which lies in the upper medulla oblongata, and by the MVN. These structures project directly to the abducens nucleus [7].

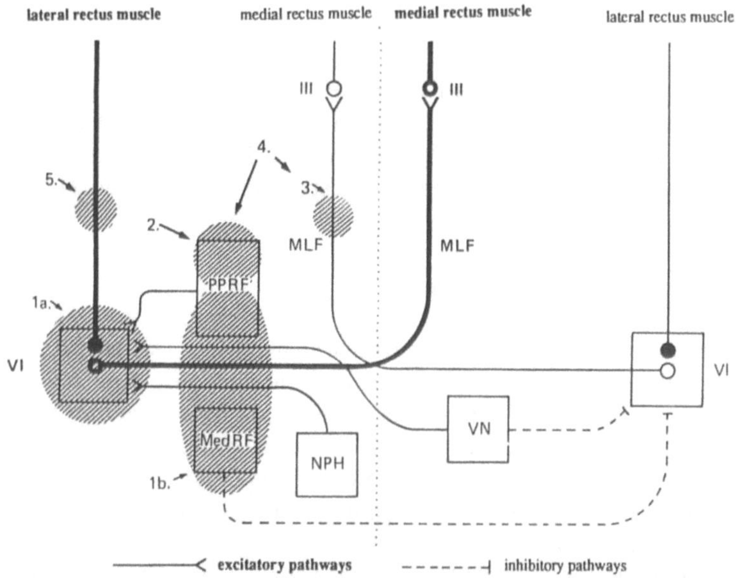

Fig. 1. The brain-stem horizontal oculomotor system. *III*, oculomotor nucleus; *VI*, abducens nucleus; *PPRF*, paramedian pontine reticular formation; *MedRF*, medullary reticular formation; *NPH*, nucleus prepositus hypoglossi; *VN*, vestibular nuclei; *MLF*, medial longitudinal fasciculus. Lesions: *1a*, *1b*, complete horizontal gaze palsy; *2*, horizontal saccade paresis; *3*, internuclear ophthalmoplegia; *4*, one-and-a-half syndrome; *5*, fascicular abducens nerve palsy

Brain-stem Lesions Affecting Horizontal Eye Movements

Complete Paresis of Horizontal Gaze

Complete paresis of horizontal gaze is usually caused by a lesion of the abducens nucleus (stage one of our case; Fig. 1, lesion 1a). Characteristically, there is concomitant infranuclear facial palsy due to involvement of the fasciculus of the seventh cranial nerve. Horizontal gaze palsy without facial paresis may result from a lesion of the caudal PPRF (Fig. 1, lesion 1b), which extends to adjacent vestibular projections to the abducens nucleus [8]. If the lesion encompasses the upper MedRF ipsiversive saccades in the intact field of gaze are slowed, since inhibition of the antagonists fails [11].

Bilateral damage to the mediocaudal PPRF results in slowing of vertical saccades [3]. The most likely explanation is involvement of PPRF neurons which project to the rostral interstitial nucleus of the MLF, a critical structure for the generation of vertical saccades [4].

Horizontal Saccade Paresis

Lesions of the medial and rostral PPRF cause either slowing or absence of ipsiversive saccades [1, 5]. Vestibular and pursuit eye movements are preserved as the vestibular projections to the abducens nucleus, which are located more caudal and lateral, are spared (Fig. 1, lesion 2).

Internuclear Ophthalmoplegia

Damage to abducens interneurons which project via the MLF to the contralateral oculomotor nucleus results in adduction paresis ipsilateral to the lesioned MLF for all types of horizontal conjugate eye movements (Fig. 1, lesion 3). Interruption of the MLF causes complete failure of adduction, while partial lesions result in slow and hypometric adducting movements. The origin of dissociated nystagmus in the contralateral eye is still obscure [8]. Activation of the medial rectus muscle during convergence is preserved unless the lesion extends to convergence neurons, which are located around the oculomotor nucleus [8].

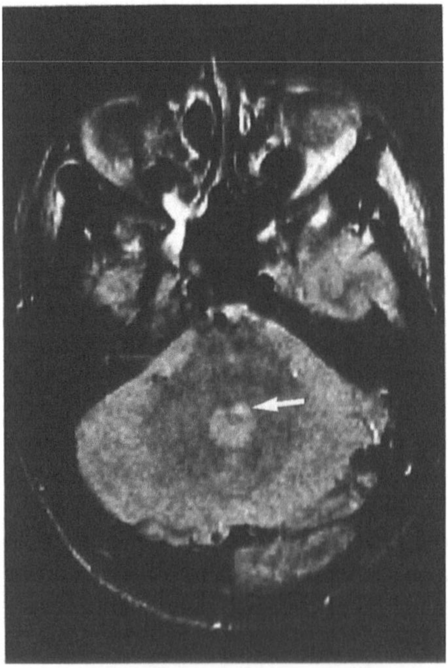

Fig. 2. T_2-weighted magnetic resonance imaging of the pons shows a small hyperintense lesion in the dorsal left paramedian tegmentum (*arrow*) with a size of 0.5 cm

One-and-a-half Syndrome

If a lesion affects the PPRF or the abducens nucleus in addition to the MLF on one side an ipsiversive horizontal gaze palsy and a contraverive adduction deficit combine as the so-called one-and-a-half syndrome (Stage two of our case; Fig. 1, lesion 4). In primary position there may be an exotropia of the contralateral eye due to unopposed drives of the unaffected abducens nucleus [12].

Fascicular Abducens Nerve Palsy

Damage to the fibers of the abducens motorneurons, which travel ventrolaterally through the lower pons, cause a typical abducens nerve palsy (Fig. 1, lesion 5). The clinical signs are horizontal diplopia, which is greatest when looking ipsilaterally with uncrossed images (stage three of our case). Differentiation from a peripheral lesion is feasible only with additional clinical or neurophysiological evidence of brain-stem dysfunction, such as contralateral hemiparesis in the Millard-Gubler syndrome.

Conclusion

Based on the brain-stem anatomy of horizontal eye movements, our patient had a lesion of the abducens nucleus extending to the seventh nerve fascicle (stage one) which subsequently expanded, thus reaching the ipsilateral MLF (stage two) and the sixth nerve fascicle (stage three). T_2 magnetic resonance imaging showed a corresponding lesion in the dorsocaudal pontine tegmentum which was most probably caused by acute demyelination (Fig. 2). All signs and symptoms resolved within 20 days.

References

1. Deleu D, Solheid C, Michotte A, Ebinger G (1988) Dissociated ipsilateral horizontal gaze palsy in one-and-a-half syndrome: a clinicopathologic study. Neurology 38:1278–1280
2. Graybiel AM, Hartwieg A (1974) Some afferent connections of the oculomotor complex in the cat: an experimental study with tracer techniques. Brain Res 81:543–549
3. Hanson MR, Hamid AM, Tomsak RL, Chou SS, Leigh RJ (1986) Selective saccadic palsy caused by pontine lesions: clinical, physiological, and pathological correlations. Ann Neurol 20:209–217
4. Henn V, Lang W, Hepp K, Reisine H (1984) Experimental gaze palsies in monkeys and their relation to human pathology. Brain 107:619–636
5. Kommerell G, Henn V, Bach M, Lücking CH (1987) Unilateral lesion of the paramedian pontine reticular formation. Neuroophthalmology 7:93–98

6. Langer T, Fuchs AF, Chubb MC, Scudder CA, Lisberger SG (1985) Floccular efferents in the rhesus macaque as revealed by autoradiography and horseradish peroxidase. J Comp Neurol 235:26–37
7. Langer T, Kaneko CRS, Scudder CA, Fuchs AF (1986) Afferents to the abducens nucleus in the monkey and cat. J Comp Neurol 245:379–400
8. Leigh RJ, Zee DS (1991) Diagnosis of central disorders of ocular motility. In: Leigh RJ, Zee DS (eds) The neurology of eye movements. Philadelphia, pp 378–530
9. McCrea RA, Strassman A, Highstein SM (1986) Morphology and physiology of abducens motoneurons and internuclear neurons intracellularly injected with horseradish peroxidase in alert squirrel monkeys. J Comp Neurol 243:293–308
10. McCrea RA, Strassman A, May E, Highstein SM (1987) Anatomical and physiological characteritics of vestibular neurons mediating the horizontal vestibulo-ocular reflex of the squirrel monkey. J Comp Neurol 264:547–570
11. Pierrot-Deseilligny O (1988) Brainstem control of horizontal gaze: effect of lesions. In: Kennard C, Clifford Rose F (eds) Physiological aspects of chinical neuro-ophthalmology. Chapman and Hall, London, pp 209–235
12. Sharpe JA, Rosenberg MA, Hoyt WF, Daroff RB (1974) Paralytic pontine exotropia. A sign of acute unilateral pontine gaze palsy and internuclear ophthalmoplegia. Neurology (Minneap) 24:1076–1081
13. Strassman A, Highstein SM, McCrea RA (1986) Anatomy and physiology of saccadic burst neurons in the alert squirrel monkey. I. Excitatory burst neurons. J Comp Neurol 249:337–357
14. Strassman A, Highstein SM, McCrea RA (1986) Anatomy and physiology of saccadic burst neurons in the alert squirrel monkey. II. Inhibitory burst neurons. J Comp Neurol 249:358–380

The Oculo-auricular Phenomenon in Brain-Stem Disease

P.P. Urban and H.C. Hopf

Neurologische Klinik, Universität Mainz, Langenbeckstraße 1, 55131 Mainz, Germany

Introduction

The oculo-auricular phenomenon (OAP), as described by Wilson in 1908 [26], consists of bilateral coactivation of external ear muscles during lateral gaze. Two ear muscles are innervated by motoneurons of the medial part of the facial nucleus [9, 25]: the transverse auricular muscle (TAM) (Fig. 1), which adducts and curls the helix of the auricle, and the posterior auricular muscle (Fig. 2), which moves the whole auricle backwards [3, 16, 22].

There is evidence that the superior colliculi are involved in this coactivation. Electrical stimulation of one superior colliculus evokes contralateral gaze deviation and bilateral pinna movements in mammals [14, 15, 18–20]. Morphologically, two projections from the superior colliculi to the facial nuclei have been identified, one to the ipsilateral and contralateral facial nuclei [1, 12, 23], the other via interneurons located in the ipsilateral paralemniscal zone [2, 5] to the contralateral facial nucleus [2, 4, 13, 21, 24]. It is still controversial whether the level of midline crossing is pontine [21] or mesencephalic [24].

Registration of the OAP

The OAP was evaluated by recording the activity of the TAM electromyographically. Bipolar surface electrodes were placed at the dorsum of the ear, close to the helix. Horizontal eye movements were recorded using electrooculography.

The Normal OAP Pattern

Here we discuss the coactivation of the TAM only. The normal pattern (Fig. 3) shows bilateral tonic activation during lateral gaze. It occurs in the majority of patients at maximal gaze deviation in saccadic as well as pursuit movements. Coactivation may be less well visualized with weak lateral eye

L.R. Caplan and H.C. Hopf (Eds.)
Brain-Stem Localization and Function

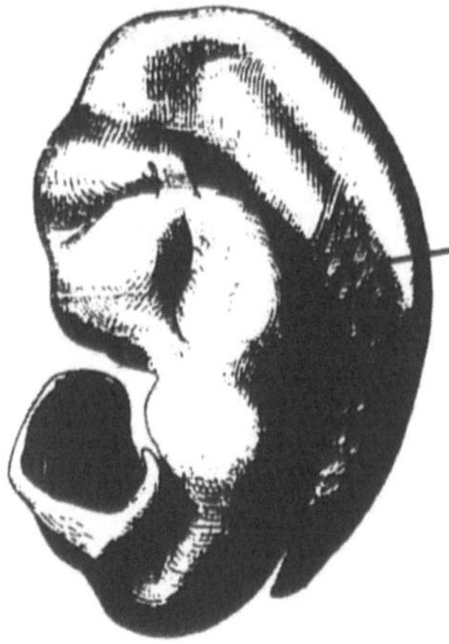

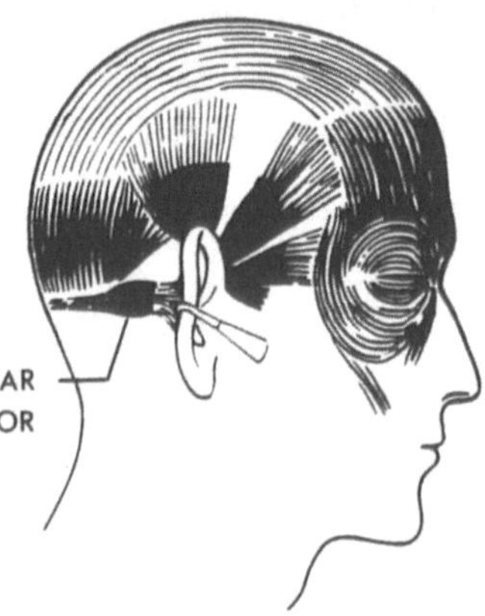

Fig. 1 (*left*). Site of the transverse auricular muscle

Fig. 2 (*right*). Site of the posterior auricular muscle

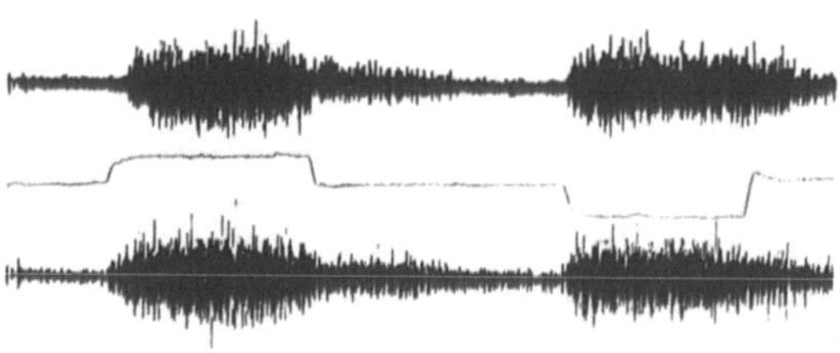

Fig. 3. Normal coactivation pattern of OAP. *Upper trace*, right TAM; *middle trace*, eye movements (*up*: maximal right gaze; *down*: maximal left gaze); *lower trace*, left TAM

deviation. In 3.9% of subjects transverse auricular muscle activity was not recordable. This is probably due to the fact that in man it is an atavistic muscle which might not be developed [17].

The Abnormal OAP Pattern

We examined 1186 patients with brain-stem lesions. The TAM showed unilateral (27 patients = 2.3%) or bilateral (19 patients = 1.6%) absence of activity. This may be due to peripheral facial palsy or atavism [17]. Thus, unilateral or bilateral absence of TAM activity is not a valid sign of a brain-stem lesion.

There are three different abnormal patterns indicating brain-stem lesions (197 patients = 16.6%) (Fig. 4):

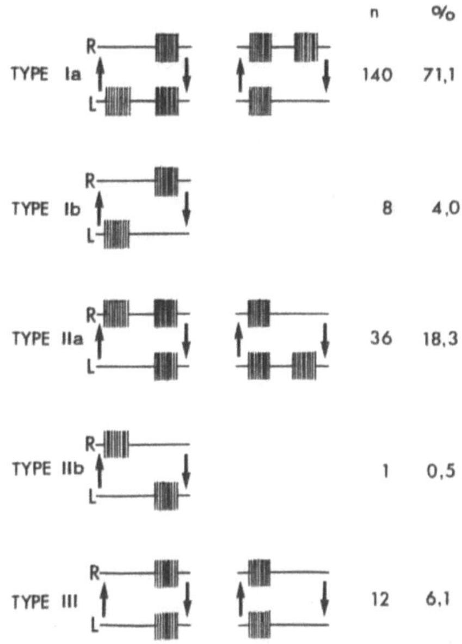

Fig. 4. Abnormal OAP patterns as observed in 197 patients with brain-stem lesions. [From Urban (1993) Brain 116:727–738]

1. Absence of coactivation homolateral to right or left gaze (type I, a) or either direction (type I, b)
2. Absence of coactivation contralateral to right or left gaze (type II, a) or either direction (type II, b)
3. Bilateral absence of coactivation with right or left gaze (type III). Here, activation of one or both TAM on moving the eyes to one side excludes peripheral causes of muscle silence on moving the eyes to the other side.

Localization of Lesions Causing Abnormal OAP Patterns

Brain-stem lesions causing OAP abnormality can be demonstrated by imaging techniques (MRI) or associated abnormality of brain-stem reflexes, such as the blink reflex (BlinkR), masseter reflex (MassR), or absence of the caloric vestibular response. Isolated unilateral changes of the BlinkR-R1 component are a valid finding with small homolateral midpontine lesions [10, 11]. Unilateral MassR abnormalities occur homolateral to small midbrain lesions [7, 8]. Impairment of the caloric response indicates a homolateral lower pontine lesion involving the vestibular nuclei if peripheral causes are excluded [6].

During the past 3 years, we carefully selected patients with small brain-stem lesions and tested the OAP, the BlinkR, the MassR, and the caloric

OAP-PATTERN	BLINK-REFLEX n=8 R	L	MASSETER-REFLEX n=22 R	L	VESTIBULAR LESION n=7 R	L
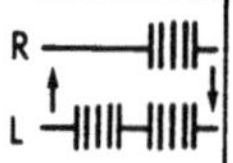	4		5	9	4	
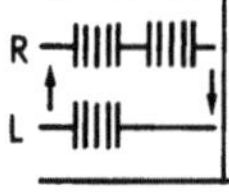		4	8			3

Fig. 5. Abnormal findings with blink reflex, masseter reflex, and caloric response in 37 patients with the indicated abnormal OAP patterns

vestibular response (Fig. 5). Absence of homolateral OAP activity (type Ia) correlates highly with an ipsilateral BlinkR-R1 abnormality (90%) and impairment of the caloric vestibular response (100%). Both conditions reliably indicate that an ipsilateral mid to lower pontine location of the lesion causes the type Ia OAP pattern.

The type Ia OAP pattern was also found to be associated with a contralateral MassR abnormality (77%). On the other hand, there were patients showing the type Ia OAP pattern and homolateral MassR and coincidental homolateral BlinkR-R1 or caloric response abnormalities (23%). Changes of the MassR in about 2/3 of patients with INO occur with midbrain lesions [8]. In the remaining 1/3, however, the location is mid to upper pontine, as identified by simultaneous MassR and BlinkR-R1 abnormalities [8]. Thus, the MassR findings support contralateral midbrain lesions in the majority of patients and homolateral pontine lesions in the residual patients. The remaining abnormal patterns of OAP activity (types Ib, II, III) were too rare to allow for conclusions regarding localization.

We conclude from our findings that the type Ia OAP pattern results from involvement of supranuclear (with respect to the facial nucleus) structures. The location of the lesion is either midpontine ipsilateral or in the midbrain contralateral. This supports the view [21] that the descending tracts cross the midline at a rostral or midpontine level. The connections between the superior colliculi and the ipsi- and contralateral facial nuclei demonstrated in amimals [1, 2, 5, 12, 23] are likely to also exist in man.

References

1. Berthoz A, Grantyn A (1986) Neuronal mechanisms underlying eye-head coordination. In: Freund HJ, Büttner U et al. (eds) The oculomotor and skeletalmotor systems: differences and similarities. Elsevier, Amsterdam, p 325

2. Henkel CK, Edwards CB (1978) The superior colliculus control of pinna movements in the cat: possible anatomical connections. J Comp Neurol 182:763–776
3. Heuser M (1976) Das Oculo-auricular-Phänomen. J Neurol 212:95–99
4. Hinrichsen CFL, Watson CD (1983) Brain stem projections to the facial nucleus of the rat. Brain Behav Evol 22:153–163
5. Holstege G, Collewijn H (1982) The efferent connections of the nucleus of the optic tract and the superior colliculus in the rabbit. J Comp Neurol 209:139–175
6. Hopf HC (1987) Vertigo and masseter paresis. J Neurol 235:42–45
7. Hopf HC, Gutmann L (1990) Diabetic 3rd nerve palsy: evidence for a mesencephalic lesion. Neurology 40:1041–1045
8. Hopf HC, Thömke F, Gutmann L (1991) Midbrain vs. pontine medial longitudinal fasciculus lesions: the utilization of masseter and blink reflexes. Muscle Nerve 14: 326–330
9. Jenny AB, Saper CB (1987) Organisation of the facial nucleus and corticofacial projection in the monkey: a reconsideration of the upper motor neuron facial palsy. Neurology 37:930–939
10. Kimura J (1983) Clinical use of the electrically elicited blink reflex. In: Desmedt JE (ed) Motor control mechanisms in health and disease. Raven, New York, pp 773–786
11. Krämer G, Lowitzsch K, Besser R, Hopf HC (1986) Isolierte R1-Veränderungen des Blinkreflexes. In: Lowitzsch K (ed) Hirnstammreflexe. Thieme, Stuttgart, pp 124–131
12. May PJ, Vidal P-P, Baker R (1986) Anatomy of the tectofacial pathways in the rat. Soc Neurosci Abstr 12:972
13. Panneton WM, Martin GF (1983) Brainstem projections to the facial nucleus of the oppossum. A study using axonal transport techniques. Brain Res 267:19–23
14. Roucoux A, Crommelinck M (1976) Eye movements evoked by superior colliculus stimulation in the alert cat. Brain Res 106:349–363
15. Schaefer KP (1970) Unit analysis and electrical stimulation in the optic tectum of rabbits and cats. Brain Behav Evol 3:222–240
16. Schmidt D, Thoden U (1978) Co-activation of the m. transversus auris with eye movements. Albrecht von Graefes Arch Ophthalmol 206:227–236
17. Schwalbe G (1897) Das äußere Ohr. In: von Bardeleben K (ed) Handbuch der Anatomie des Menschen, vol 5/2. Fischer, Jena, pp 154–155
18. Stein BE (1984) Multimodal representation in the superior colliculus and optic tectum. In: Vanegas H (ed) Comparative neurology of the optic tectum. Plenum, New York, p 819
19. Stein BE, Clamann HP (1981) Control of pinna movements and sensorimotor register in cat superior colliculum. Brain Behav Evol 19:180–192
20. Syka J, Radil-Weiss T (1971) Electrical stimulation of the tectum in freely moving cats. Brain Res 28:567–572
21. Takeuchi Y, Nakano K, Uemura M et al. (1979) Mesencephalic and pontine afferent fibre system to the facial nucleus in the cat: a study using horseradish peroxidase and silver impregnation techniques. Exp Neurol 66:330–342
22. Urban P (1988) Aktivierungscharakteristik des atavistischen M. auricularis posterior. Dissertation, University of Mainz
23. Vidal PP, May PJ, Baker R (1985) Electrophysiology of the tectofacial pathway. Soc Neurosci Abstr 11:82
24. Vidal PP, May PJ, Baker R (1988) Synergistic organization of the tectal-facial pathways in the cat. J Neurophysiol 60:769–797
25. Welt C, Abbs JH (1990) Musculotopic organization of the facial motor nucleus in *Maccaca* fascicularis: a morphometric and retrograde tracing study with cholera toxin B-HRP. J Comp Neurol 291:621–636
26. Wilson SAK (1908) A note on an associated movement of the eyes and ears in man. Rev Neurol Psychiatry 6:331–336

Sensory Functions and Trigeminal Reflexes

Correlation of Somatosensory Evoked Potentials and Somatosensory Findings in Patients with Brain-Stem Lesions

R. Claß and U.W. Buettner

Neurologische Klinik, Kliniken Schnarrenberg, Eberhard Karls-Universität, Hoppe-Seyler-Straße 3, 72076 Tübingen, Germany

Introduction

There are many reports of poor correlation between clinical findings and somatosensory evoked potentials (SEP). This holds especially for patients with demyelinating disease, where pathological latency delays and amplitude reductions are common even in patients without sensory deficits or specific complaints. Until now there has been no sufficiently accepted explanation for this phenomenon [2, 3, 4]. The notion of the high sensitivity and low specificity of the SEP triggered the present study; therefore, we attempted to correlate somatosensory and SEP abnormalities and to provide new data to localize brain-stem lesions by means of the SEP.

Methods

Thirteen patients with lesions confirmed by computed tomography (CT) or magnetic resonance imaging (MRI) gave informed consent to an examination of all somatosensory modalities (cold, warm, pain, light touch, vibration, position sense, kinesthesia, and two-point discrimination). The respective SEP were recorded from the upper or lower limbs. The latter recording was performed according to accepted methodological standards. SEP were obtained following electrical stimulation of the nerve with 0.1-ms square wave pulses 4 mA above the motor threshold. Recordings were made using platinum needle electrodes inserted into the scalp above the sensory representation of the limb and a Fz reference electrode. Additional recording sites were Erb's point and above the spinal process C7. In addition, C'_3/C'_4 respectively were referenced against the knee. The bandpasses of the amplifiers were set at 10–1000 Hz. All data were processed conventionally by a four-channel averager (Toennies DA4).

L.R. Caplan and H.C. Hopf (Eds.)
Brain-Stem Localization and Function

Results

Seven out of 13 patients had ischemic lesions, four of them in the vascular region causing a lateral medullary syndrome, one causing a lateral thalamic syndrome, and two in the upper and paramedian pons. The other lesions were two bleedings (tectum and pons), two angiomas (pons), an epidermoid tumor in the left cerebellar pontine angle, and an astrocytoma in the pontomesencephalic transition. The overall results showed eight normal SEP among all patients with the lateral medullary syndrome. In all patients with exclusive involvement of the spinothalamic system, the SEP were normal. Additional lemniscal symptoms (two-point discrimination, position sense, vibration) led in three of four patients to pathological SEP changes. There was no single somatosensory predictor for SEP pathology. In one patient with severe loss of position sense, mild two-point discrimination deficit, and paresthesias, the SEP was normal.

Case Report

The patient was a 51-year-old man with hypesthesia and hypalgesia of the left side and bilateral reduction of two-point discrimination. Position sense and vibration were normal. MRI showed a pontine cavernoma (Fig. 1) at the border of the tegmentum and the right paramedian basis pontis. The extent of the lesion corresponds exactly to that of the right medial lemniscus and encroaches on the trigeminospinothalamic and the spinothalamic tracts. Brain-stem auditory potentials, trigeminal, and tibial nerve SEP were ab-

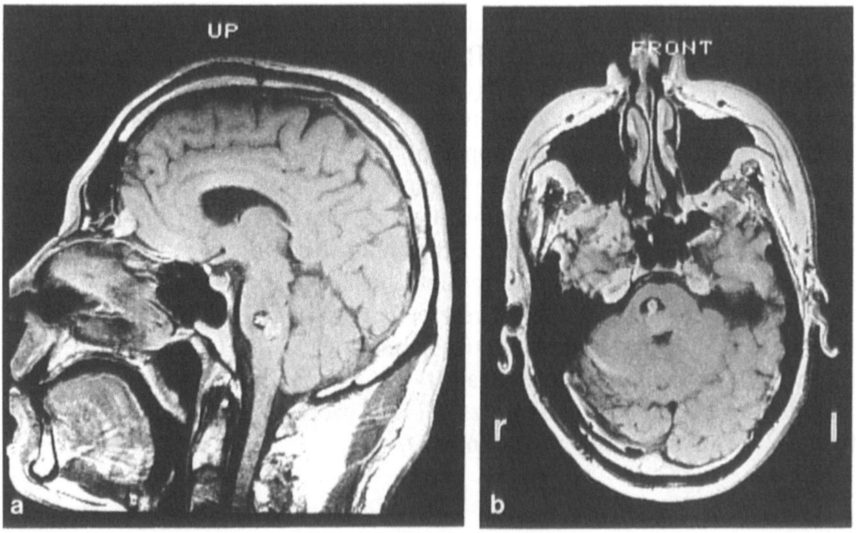

Fig. 1a,b. Magnetic resonance imaging of a 51-year-old patient with a pontine cavernoma (see text)

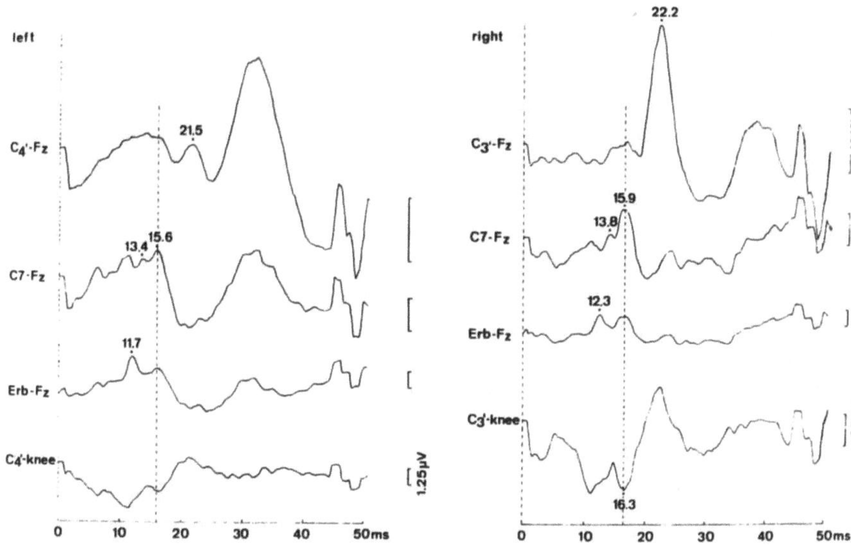

Fig. 2. Median nerve somatosensory evoked potentials of a 51-year-old patient with a pontine cavernoma. It shows intact subcortical far-field potentials up to P14 marked by the cursor in the bottom recordings. The latency corresponds to N14 recorded at C7. The actual latency is around 16 ms. The main abnormality is an amplitude reduction of N20 following left stimulation. Latencies are normal

normal. Only the median nerve SEP is shown because of its peculiar finding (Fig. 2); it documents normal subcortical farfield potentials P14 on both sides, whereas the cortical N20 is reduced in amplitude on the left side.

Discussion

There are two aspects of major interest in this study. First, there is the question of the generation of evoked potentials in the brain-stem. For the median nerve SEP it is accepted that P14 is a subcortically generated far-field potential when SEP is recorded from the scalp with a noncephalic reference electrode [5, 8, 10, 11]. This is clearly demonstrated in the patient with a pontine cavernoma (Fig. 1), who shows a preserved P14, but abnormal N20, providing further proof of the origin of P14 being caudal to the mid-pontine brain-stem. Any attempts at understanding why only two-point discrimination is affected, but not position sense and vibration sensation, remain speculative.

The second and main aspect of this study attempts to correlate clinical (somatosensory) data to electrical activity of the brain-stem (SEP) in patients with known lesion sites. The anatomy reveals a peculiar course of the main sensory afferent pathways, mainly the medial lemniscal and the spinothalamic pathways. Caudal to the pons, the pathways are separated from each other, but more rostrally both pathways join and follow an adjacent course. This means that, from an anatomical point of view, it may be difficult to find lesions that differentiate between modalities conveyed by the lemniscal or spinothalamic pathways. However, it is reasonable to assume that in patients with dorsolateral medullary infarctions (Wallenberg syndrome), lemniscal function is spared and, therefore, according to our present understanding, SEP is unaffected [6, 7]. On the other hand, more rostrally located lesions eventually affect both pathways together, which should elicit more consistent correlations between sensory disturbances and SEP. This hypothesis remains to be tested in future research. Our preliminary data suggest more complicated relations between SEP and sensation, since although position sense is widely accepted as a marker modality for the SEP, there are many exceptions to this rule [1].

Acknowledgements. The help of Mrs. H. Uhl and Mrs. C. Winkler during recordings and preparation of the figures is gratefully appreciated.

References

1. Anziska B, Cracco RQ (1980) Short latency somatosensory evoked potentials: studies in patients with focal neurological disease. Electroencephalogr. Clin Neurophysiol 49:227–239
2. Buettner UW (1991) Critical evaluation of somatosensory evoked potentials in multiple sclerosis. In: Wiethölter H, Dichgans J, Mertin J (eds) Current concepts in multiple sclerosis. Excerpta Medica, Elsevier, Amsterdam, pp 61–66
3. Chiappa KH (1990) Evoked potentials in clinical medicine. Raven Press, New York
4. Chiappa KH, Choi S, Young RR (1980) Short latency somatosensory evoked potentials following median nerve stimulation in patients with neurological lesions. In: Desmedt JE (ed) Progress in clinical neurophysiology, vol 7, Karger, Basel, pp 264–281
5. Dillmann U, Besser R, Eghbal R, Koehler J, Ludwig B (1990) SEP and MRI findings in patients with localized brain-stem lesions. In: P.M. Rossini, F. Mauguiere (eds) New trends and advanced techniques in clinical neurophysiology, EEG Suppl. 41. Elsevier, Amsterdam, pp 314–319
6. Halliday AM, Wakefield GW (1963) Cerebral evoked potentials in patients with dissociated sensory loss. J Neurol Neurosurg Psychiatry 26:211–219
7. Holmgren H, Leijon G, Boivie J, Johansson I, Ilievska L (1990) Central post-stroke pain – somatosensory evoked potentials in relation to location of the lesion and sensory signs. Pain 40:43–52
8. Mauguière F, Brunon AM, Echallier JF, Courjon J (1982) Early somatosensory evoked potentials in lesions of the lemniscal pathway in humans. In: Courjon J, Mauguière F, Revol M (eds) Clinical applications of evoked potentials in neurology. Raven Press, New York, pp 321–338
9. Sawada H, Seriu N, Udaka F, Kameyama M (1990) Magnetic resonance imaging of medial medullary infarction. Stroke 21:963–966
10. Tomberg C, Desmedt JE, Ozaki I, Noël P (1991) Nasopharyngeal recordings of somatosensory evoked potentials document the medullary origin of the N18 far-field. Electroencephalogr. Clin Neurophysiol 80:496–503
11. Yamada T, Kayamori R, Kimura J, Beck DO (1984) Topopgraphy of somatosensory evoked potentials after stimulation of the median nerve. Electroencephalogr. Clin Neurophysiol 59:29–43

Functional Anatomy of the Spinal Trigeminal Nucleus

H. Schröder

Institut für Physiologische Chemie und Pathobiochemie,
Johannes Gutenberg-Universität, Duesbergweg 6, 55128 Mainz, Germany

Introduction

The trigeminal nerve (nV) conveys general somatic afferent exteroceptive input from the skin of the face and the frontal parts of the head, the mucous membranes of the oral cavity, tongue and nose, the tooth pulp, eye and meninges via the trigeminal (Gasserian) ganglion to the nuclei of termination [28]. Somatosensory input reaches the principal sensory nucleus of the nV and the spinal trigeminal nucleus of the nV (STN). Proprioceptive input from masticatory muscles innervated by trigeminal motor efferents is transferred to the nV mesencephalic nucleus.

The STN, its afferent and efferent connections, are of special clinical importance because of their involvement in the processing of nociceptive and thermal input from cutaneous regions of the face [10]. In man, lesions of the STN result in decreased sensitivity to painful stimuli and temperature applied to the face. On the experimental level, these observations have been corroborated neurophysiologically in several species [7]. The STN medullary tract as a target site for the surgical treatment of trigeminal neuralgia underlines the importance of the STN in pain perception and management. This paper summarizes the functionally most important anatomical data on the STN. Since most of the relevant work has been performed on experimental animals, such findings are the basis for much of the information. Wherever possible I have included data on the human STN.

Subnuclei of the Spinal Trigeminal Nucleus

Phylogenetic, ontogenetic, and cytoarchitectonic studies have shown the STN of primates – including man [21] – and higher vertebrates to be composed of three different subnuclei in a rostrocaudal direction: the subnucleus oralis (SoSTN), the subnucleus interpolaris (SiSTN), and the subnucleus caudalis (ScSTN) (Fig. 1). Descriptions of the cytoarchitectonics of the subnuclei are beyond the scope of this review. The interested reader is referred to the references.

L.R. Caplan and H.C. Hopf (Eds.)
Brain-Stem Localization and Function

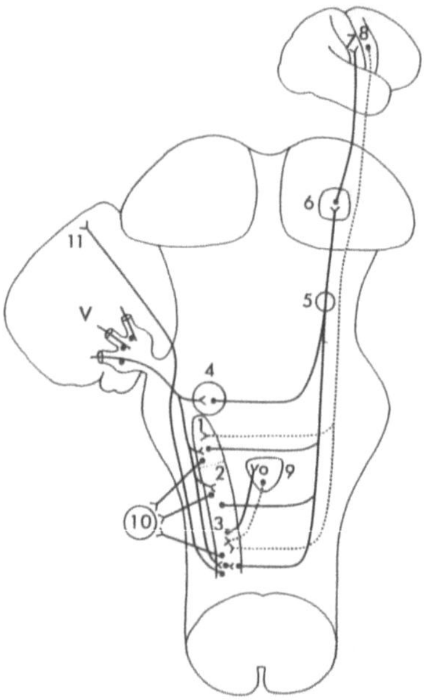

Fig. 1. The most important afferent and efferent connections of the spinal trigeminal nucleus (STN). *1*, subnucleus oralis (So) of the STN; *2*, subnucleus interpolaris (Si) of the STN; *3*, subnucleus caudalis (Sc) of the STN; *4*, principal nucleus of the fifth nerve; *5*, lemniscus trigeminalis; *6*, thalamic nuclei representing parts of the posteromedial ventral, the posterolateral ventral, the ventral lateral, and the intralaminar nuclei; *7*, somatosensory cortex; *8*, areae 4, 5, and 6 (Brodman); *9*, periaqueductal gray matter; *10*, interneuronal connections to the trigeminal and hypoglossal motor nuclei; *11*, cerebellar cortex. (Modified from [19])

The Subnucleus oralis

The SoSTN extends caudally from the principal sensory nucleus of the nV (level of the second part of the root of the facial nerve) to the level of the oral third of the inferior olive. Only one rather small neuronal cell type is encountered in the SoSTN [21].

The Subnucleus interpolaris

The SiSTN – caudally delineated from the SoSTN by the obex – is a uniform area containing small and large neurons loosely packed among numerous fascicles of fibers lacking the typical lamination displayed by the ScSTN [21, 22].

The Subnucleus caudalis

The ScSTN extends caudally to the level just above the decussation of the pyramids, where it continues into the dorsal horn of the cervical spinal cord [21]. From lateral to medial the ScSTN can be further divided into three different zones: (a) the marginal zone, (b) the substantia gelatinosa, and (c) the nucleus proprius [21]. This arrangement closely resembles that of the dorsal horn of the spinal cord, which has similar structures for afferent connections [19, 26].

Afferent Connections of the Spinal Trigeminal Nucleus

Peripheral Afferent Connections

Long descending branches of the afferent trigeminal root form the spinal trigeminal tract (STT), which terminates in a somatotopic fashion in the STN. It is widely agreed that mandibular (V_3), maxillar (V_2), and ophthalmic (V_1) divisions of the nV are represented in a dorsoventral sequence within the STT and the STN. Electrical stimulation of nV branches elicits responses in the STN [31]. Detailed data on STN somatotopy of different primary afferents are available for the rat and the cat.

Studies on the topographic organization of V_3 afferents in rat STN have been performed by Takemura et al. [29]. In the SoSTN auriculotemporal and mylohyoid nerve (cutaneous branch) input has been shown to be located at the dorsolateral border of the mandibular division, while inferior alveolar nerve endings were located medial to that of the auriculotemporal and mylohoid nerve in the dorsal two thirds of the mandibular division. In the SiSTN auriculotemporal and mylohyoid afferents show the same arrangement as in the SoSTN, while the inferior alveolar nerve terminates in the medial half of the mandibular division ventral to the mylohyoid terminal field. In general primary neurons innervating intraoral structures terminate medial (SoSTN) and ventral (SiSTN) to those innervating the facial skin.

Shigenaga et al. [25] have shown the subnucleus-specific representation of cat nV-innervated facial and oral regions. Afferents from the latter project to the ventral and dorsal subnuclei of the principal nV nucleus, the rostrodorsomedial and dorsomedial regions of the SoSTN, and the dorsomedial part of the SiSTN [1]. These findings correspond with clinical observations of perioral, endoral, and lingual disturbances of sensitivity caused by rostral STN lesions (Hopf, personal communication).

The anterior facial skin – innervated by mental, infraorbital, and frontal nerves – is represented in the ventral or ventrolateral regions of each subnucleus. Afferents that innervate the peripheral or posterior facial skin project mainly to the rostrocaudal midlevels of the subnucleus ventralis of the principal nucleus and the caudal parts of the SiSTN. Corneal afferents terminate in the caudal two-thirds of the subnucleus ventralis of the principal nucleus and the caudal third of the SiSTN. Taken together, a clear somatotopic organization with the face inverted is represented in the ventral subnucleus of the principal nucleus and the caudal SiSTN.

Electrophysiologically, in the SiSTN Hayashi et al. [16] confirmed the medially facing, inverted-head representation characteristic of other nV sensory nuclei; i.e., intraoral receptive fields (RF) are represented medially, that of the lateral face laterally, the V_3 RF is located dorsomedially, the V_1 one ventrolaterally and the maxillary RF between that of V_3 and V_1. SiSTN nociceptive neurons projecting to the thalamus (see below) may contribute to the partial orofacial pain sensitivity reported to remain after trigeminal tractotomy.

Primary afferents within the substantia gelatinosa of the cat ScSTN have been shown to be arranged in an onionskin-like fashion, indicating the ScSTN to be a morphologically and functionally different unit compared with the other nV sensory nuclei [24]. This is of clinical interest, since lesions of the STT and/or the STN in man cause onionskin-like hypesthetic facial areas confined by the so-called lines of Sölder, which may be related to the topical arrangement of RFs in the ScSTN.

Ascending and descending internuclear projections within the cat trigeminal sensory complex have been studied by Nasution and Shigenaga [18]. Main ascending fibers arise from the ScSTN and the caudal SiSTN. The former project to the rostral nV nuclei – except the caudal SiSTN and the dorsal principal nucleus – the latter mainly to the dorsomedial SoSTN. Descending axons originate in all sensory nV nuclei targeting at neighboring nuclei, while the caudal SiSTN and ScSTN superficial marginal zone do not receive descending input.

Obviously, internuclear connections have little impact on trigeminothalamic projections, since SoSTN neurons do not project to the thalamus.

Another issue of great functional importance is the nuclear representation of different types of primary afferents which has been studied in cat STN [15]. Functionally identified cutaneous primary afferents were labeled intraaxonally and their terminal fields were determined. Large myelinated primary afferents (i.e., input from vibrissae, G_1-hair, and slowly adapting type-I afferents) have collateral terminal arbors in the outer part of the STN near the obex, in the SoSTN, and in the marginal zone of the ScSTN. Among the small myelinated primary afferents, D-hair fibers show a similar distribution, whereas the nociceptive high-threshold mechanoreceptive afferents terminate exclusively in the superficial layers of SiSTN and ScSTH. Obviously, there are large differences in the terminations of nociceptive and non-nociceptive primary afferents, the SiSTN receiving direct nociceptive input from the facial skin.

Chemoanatomical features of the STN are beyond the scope of this paper. Some data, however, will be provided on enkephalin-containing structures, as supranuclear descending as well as peripheral input may influence enkephalinergic STN nociceptive processing.

Enkephalin-containing neurons in the rat ScSTN superficial marginal zone have been implicated in a pain-controlling circuitry. Preproenkephalin A gene expression is increased in those neurons upon electrical stimulation of nV afferents. The expression of mRNA corresponding to a major pain-modulating transmitter gene [20] is a good example of input-related plasticity at the single-neuron level.

Central Afferent Connections

Cerebral Cortex. Axons originating in the primary and secondary somatosensory areas, as well as those from Brodmann's areas A4, A5, and A6, project to the contralateral sensory nuclei of the trigeminal complex [19]. In the ScSTN fibers terminate both in the deep layers and the marginal zone or in the substantia gelatinosa. By means of electrophysiology, Sunada et al. [27] have shown that the cat contralateral primary somatosensory cortex exerts a mainly inhibitory influence on SoSTN neurons in the cat providing corticofugal descending inhibition of sensory transmission in the STN.

Raphe Nuclei. Serotoninergic and enkephalinergic axons of the raphe nuclei (RN) project bilaterally to the marginal layer and the substantia gelatinosa of the ScSTN [2].

Reticular Formation. Projections of the reticular formation terminate in the margimal layer and the substantia gelatinosa of the ScSTN [3]. Those fibers have been shown electrophysiologically to inhibit nociceptive as well as non-nociceptive units [12], most likely belonging to descending pathways from the central gray matter of the mesencephalon, mediating analgesia produced by stimulation or opiates [3].

Efferent Connections of the STN

Efferents to Cranial Nerve Nuclei

Mostly ipsilateral projections reach the facial nucleus, which also receives collateral input from STN efferents to abducens and trigeminal motor nuclei [8, 11, 26, 30]. This pathway has been shown to be of major importance in mediating the blink reflex and proprioceptive motor control of facial muscles [23]. Direct ipsilateral projections from the STN to the hypoglossal nucleus have been shown in the cat [26] and in the rat [5]. Furthermore, projections to the XI, X, and IX cranial nerve nuclei are present [26].

Mesencephalon

Somatosensory projections to the mesencephalon have been studied in the monkey [32]. Using tracing studies, the SiSTN has been shown to project mainly to the superior colliculus (SC) and the intercollicular nucleus. Efferents of the ScSTN mainly reach the parabrachial nucleus and the periaqueductal gray (PAG). The multimodal input to the midbrain (auditory afferents to inferior, visual, and somatosensory afferents to superior colliculi; transmission of low-threschold cutaneous stimuli via the intercollicular nucleus) probably provides the basis for orientational behavior mediated by interactions between the different modalities.

Most likely, at least some of the fibers from the ScSTN to the PAG carry nociceptive information [32]. Among other functions, the PAG is part of an endorphin-mediated pain-suppression system [13]. The noxious input suggests a negative feedback loop, effective in suppressing pain. However, afferent STN fibers terminate in the lateral part of the PAG, while efferent fibers to the STN originate in the ventral part. The existence of interneurons has not been demonstrated so far.

Thalamus

ScSTN efferents representing protopathic ascending pathways form the lateral trigeminothalamic tract that crosses to the contralateral side at the level of the medulla oblongata. The cells of origin are located in the marginal zone of the contralateral ScSTN [17]. Terminals of ascending axons are located mainly in different parts of the posterior nuclear groups of the contralateral thalamus. These include the posteromedial ventral nucleus, the adjoining part of the posterolateral ventral nucleus, the border region of these nuclei with the ventral lateral nucleus, the posterior group, and some of the intralaminar nuclei [4, 6, 9, 14].

Corpus Striatum

Using antero- and retrograde transport techniques, ScSTN efferents to the striatum were shown [33]. Neurons in the ScSTN superficial marginal zone project contralaterally, mainly to the dorsal putamen and the ventrolateral caudate nucleus.

Cerebral Cortex

Bilateral, predominantly ipsilateral projections, mainly to layer-V and -III cortical neurons, have been shown in the cat by means of anterograde transport [33]. They were found in the orbital, sigmoid, and coronal gyri and in the cortex covering the medial and lateral banks of the presylvian sulcus.

Cerebellum

Ipsilateral projections from the ScSTN to the cat cerebellar cortex have been described [17].

Spinal Cord

Ipsilateral projections from all STN subnuclei with a preponderance of the ScSTN to the spinal cord been have been shown in the cat [17].

Conclusions

The following statements today can be considered well established:

1. The STN is an important relay center for the processing of noxious, but also non-nociceptive, input from the different nV branches.
2. The terminal fields of the latter are arranged in a somatopic fashion with a rostromedial representation of intra- and perioral regions, while the anterior facial skin is represented more ventrally and the lateral parts are located laterally. An onionskin-like arrangement is found in the caudal subnucleus of the STN.
3. The STN is interconnected with a variety of brain-stem, mesencephalic, and diencephalic nuclei and with the cerebral cortex. Of special importance in nociceptive processing are the interrelationships of STN enkephalinergic structures with the mesencephalic periaqueductal gray and with the facial nucleus subserving the blink reflex.

Most of the current knowledge is based on animal studies. Clinical observations suggest similar arrangements in the human brain. Increased efforts using modern neuroanatomical techniques for the investigation of human autopsy brains will be necessary to increase our knowledge of human STN connectivity.

References

1. Azerad J, Woda A, Albe-Fessard D (1982) Physiological properties of neurons in different parts of the cat trigeminal sensory complex. Brain Res 246:7–21
2. Basbaum AI, Fields HL (1984) Endogenous pain control systems: brainstem spinal and endorphin circuitry. Annu Rev Neurosci 7:309–338

3. Beitz AJ, Wells WE, Shepard RD (1983) The loction of brainstem neurons which project bilaterally to the spinal trigeminal nuclei as demonstrated by the double fluorescent retrograde tracer technique. Brain Res 258:305–312
4. Berkley KJ (1980) Spatial relationships between the terminations of somatic sensory and motor pathways in the rostral brain stem of cats and monkdeys. I. Ascending somatic sensory input to the lateral diencephalon. J Comp Neurol 193:283–317
5. Borke R, Nau EM (1987) The ultrastructural morphology and distribution of trigemino-hypoglosal connections labeled with horseradish peroxidase. Brain Res 422:235–241
6. Burton H, Craig AD, Poulos DA, Holt JT (1979) Efferent projections from temperature-sensitive recording loci within the marginal zone of the nucleus caudalis of the spinal trigeminal complex in the cat. J Comp Neurol 183:753–778
7. Cahusac PMB, Morris R, Salt TE, Hill RG (1990) Sensory responses of caudal trigeminal neurons to thermal and mechanical stimuli and their behavioural correlates in the rat. Neuroscience 36:543–551
8. Carpenter MB, Hanna GR (1961) Fiber projection from the spinal trigeminal nucleus in the cat. J Comp Neurol 117:117–125
9. Catsman-Berrevoets CE, Kuypers HGJM (1981) A search for corticospinal collaterals to thalamus and mesencephalon by means of multiple retrograde fluorescent tracers in cat and rat. Brain Res 218:15–33
10. Dubner R, Sessle BJ, Storey AT (1978) The neural basis of orla and facial function. Plenum, New York
11. Durand J, Gigan P, Guéritaud JP, Horcholle-Bossavit G, Tyc-Dumont S (1983) Morphological and electrophysiological properties of trigeminal neurones projecting to the accessory abducens nucleus of the cat. Exp Brain Res 53:118–128
12. Eickhoff R, Handwerker HO, McQueen DS, Schick E (1978) Noxious and tactile input to medial structures of midbrain and pons in the rat. Pain 5:99–113
13. Fields HL (1981) An endorphin-mediated analgesia system: experimental and clincal observations. Adv Biochem Psychopharmacol 28:199–212
14. Ganchrow D (1978) Intratrigeminal and thalamic projections of mucleus caudalis in the squirrel monkey (Saimiri sciureus): a degeneration and autoradiographic study. J Comp Neurol 178:281–312
15. Hayashi H (1985) Morphology of terminations of small and large myelinated trigeminal primary afferent fibers in the cat. J Comp Neurol 240:71–89
16. Hayashi H, Sumino R, Sessle BJ (1984) Functional organization of trigeminal subnucleus interpolaris: nociceptive and innocuous afferent inputs, projections to thalamus, cerebellum, and spinal cord, and descending modulation from periaqueductal gray. J Neurophysiol 51:890–905
17. Matsushita M, Ikeda M, Okado N (1982) The cells of origin of the trigeminothalamic, trigeminospinal amd trigeminocerebellar projections in the cat. Neuroscience 7:1439–1454
18. Nasution ID, Shigenaga Y (1987) Ascending and descending internuclear projections within the trigeminal sensory nuclear complex. Brain Res 425:234–247
19. Nieuwenhuys R, Voogd J, van Huijzen C (1988) The human central nervous system, 3rd edn. Springer, Berlin Heidelberg New York
20. Nishimori T, Buzzi MG, Moskowitz MA, Uhl GR (1989) Preproenkephalin mRNA expression in nucleus caudalis neurons is enhanced by trigeminal stimulation. Mol Brain Res 6:203–210
21. Olszewski J (1950) On the antomical and functional organization of the spinal trigeminal nucleus. J Comp Neurol 92:401–413
22. Olszewski J, Baxter D (1954) Cytoarchitecture of the human brain stem. Karger, Basel
23. Schröder H (1993) The facial nerve – peripheral and central connections of proprioception. Eur Arch Otolaryngol (in press)
24. Shigenaga Y, Chen IC, Suemune S, Nishimori T, Nasution ID, Yoshida A, Sato H, Okamoto T, Sera M, Hosoi M (1986) Oral and facial representations within the medullary and upper cervical dorsal horns in the cat. J Comp Neurol 243:388–408

25. Shigenaga Y, Okamoto T, Nishimori T, Suemune S, Nasution ID, Chen IC, Tsuru K, Yoshida A, Tabuchi K, Hosoi M, Tsuru H (1986) Oral and facial representations in the trigeminal principal and rostral spinal nuclei of the cat. J Comp Neurol 244:1–18
26. Stewart WA, King RB (1963) Fiber projections from the nucleus caudalis of the spinal trigeminal nucleus. J Comp Neurol 121:271–286
27. Sunada T, Kurasawa I, Hirose Y, Nakamura Y (1990) Intracellular response properties of neurons in the spinal trigeminal nucleus to peripheral and cortical stimulation in the cat. Brain Res 514:189–197
28. Székely G (1989) Cranial nerves. In: Adelman G (ed) Neuroscience year. Birkhäuser, Boston (Encyclopedia of neuroscience, suppl 1)
29. Takemura M, Sugimoto T, Sakai A (1987) Topographic organization of central terminal region of different sensory branches of rat mandibular nerve. Exp Neurol 96:540–557
30. Takeuchi Y, Nakano K, Uemura M, Matsura K, Matsushima R, Mizuno N (1979) Mesencephalic and pontine afferent fiber system to the facial nucleus in the cat: a study using horseradish peroxidase and silver impregnation techniques. Exp Neurol 66:330–342
31. Tamai Y, Iwamoto M, Tsujimoto T (1986) Pathway of the blink reflex in the brainstem of the cat: interneurons between the trigeminal and the facial nucleus. Brain Res 380:19–25
32. Wiberg M, Westman J, Blomqvist A (1987) Somatosensory projection to the mesencephalon: an anatomical study in the monkey. J Comp Neurol 264:92–117
33. Yasui Y, Itoh K, Mizuno N (1987) Direct projections from the caudal spinal trigeminal nucleus to the striatum in the cat. Brain Res 408:334–338

Somatosensory Evoked Potentials After Trigeminal Stimulation in the Diagnosis of Brain-Stem Lesions

M. Stöhr

Klinik für Neurologie und Klinische Neurophysiologie, Zentralklinikum
Krankenhauszweckverband Augsburg, Stenglinstraße 15, 86156 Augsburg, Germany

Introduction

Registrations of somatosensory evoked potentials after trigeminal stimulation (trigeminal SEP) provide an objective test of trigeminal nerve function and its central connections. Important parts of this system are located in the brain stem. The impulse wave induced after stimulation of terminal trigeminal branches passes via the trigeminal nerve to the main sensory nucleus (which corresponds functionally to the dorsal column nuclei). After decussation in the trigeminal loop, it passes to the ventrobasal nucleus of the thalamus (VPM), and finally to the primary sensory cortex [3].

Method of Investigation

The technique for investigation of trigeminal SEP varies, in particular with regard to the site of stimulation. Drechsler [4] recommends stimulation of the infraorbital or mental nerve. Bennet and Jannetta [2] stimulate the gingiva, and Altenmüller et al. [1] stimulate the edge of the tongue. In our own laboratory, stimulation is carried out on the lips, since besides the tip of the tongue these have highest innervation density [8]. Depending on the clinical symptoms and the parameters to be investigated, the upper and lower lips can be stimulated simultaneously. This has the advantage of a higher amplitude of the cortical response. However, selective upper or lower lip stimulation is also possible. In the normal patient, a triphasic cortical response is obtained with the component N13, P19, and N27. Because of its constancy, the first positive deflection is the most suitable for measurement of latency [8] (Fig. 1).

As already observed by Ishiko et al. [6], after mechanical stimulation of the face and tongue, stimulus responses are found to be spread over wide areas of the scalp. However, these have an amplitude maximum over the foot of the contralateral post-central gyrus, so that routine registration via the EEG positions C5 and C6 is recommended (Fig. 2).

L.R. Caplan and H.C. Hopf (Eds.)
Brain-Stem Localization and Function

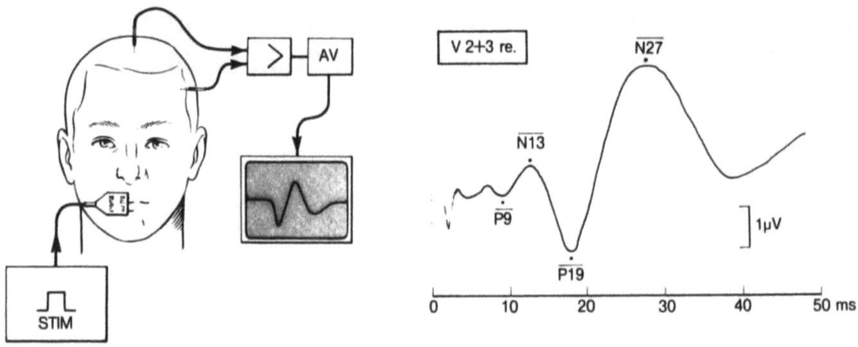

Fig. 1. *Left*: Technique for registration of the cortical response after trigeminal stimulation in the mouth (V2 and 3). *Right*: Cortical response after simultaneous stimulation of the upper and lower lips; registration from C6 against a fronto-median reference

In the context of diagnosis of trigeminal lesions, it should be borne in mind that registrations of the trigeminal SEP complement registration of the blink reflex. The blink reflex is a test of the function of the first branch of the trigeminal nerve and its connections, whereas the trigeminal SEP is a test of function of the second and third branch of the trigeminal nerve. Accordingly, different defect patterns are manifested depending on the nature of the lesion. In idiopathic trigeminal neuralgia, which preferentially involves the second and third branches, pathologically delayed trigeminal SEP are to be found on the affected side in just one half of the patients [9], whereas the blink reflex results are, as a rule, normal. Conversely, for example, surgical or traumatic trigeminal lesions which affect only the first branch or which preferentially affect this branch are accompanied by normal or only slightly pathological trigeminal SEP and simultaneous loss of the blink reflex (Fig. 3). Investigations of the blink reflex and the trigeminal SEP are thus complementary with regard to tests of the function of different parts of the trigeminal system.

A possibility of error in investigation of the trigeminal SEP which we have so far observed exclusively in patients with hemifacial spasm consists in the occurrence of abnormally high stimulus responses in the same latency range as the primary cortical potential. This is likely to be a myogenic reflex response, i.e., a disinhibited trigemino-facial reflex.

Trigeminal SEP Findings in Tumors, Infarcts, or Hemorrhages in the Brain-Stem

Typically, brain-stem lesions of varying etiology that involve the main trigeminal nucleus or the trigeminal loop lead to a reduction in amplitude on the affected side but normal or marginally prolonged latency. A P19 latency of

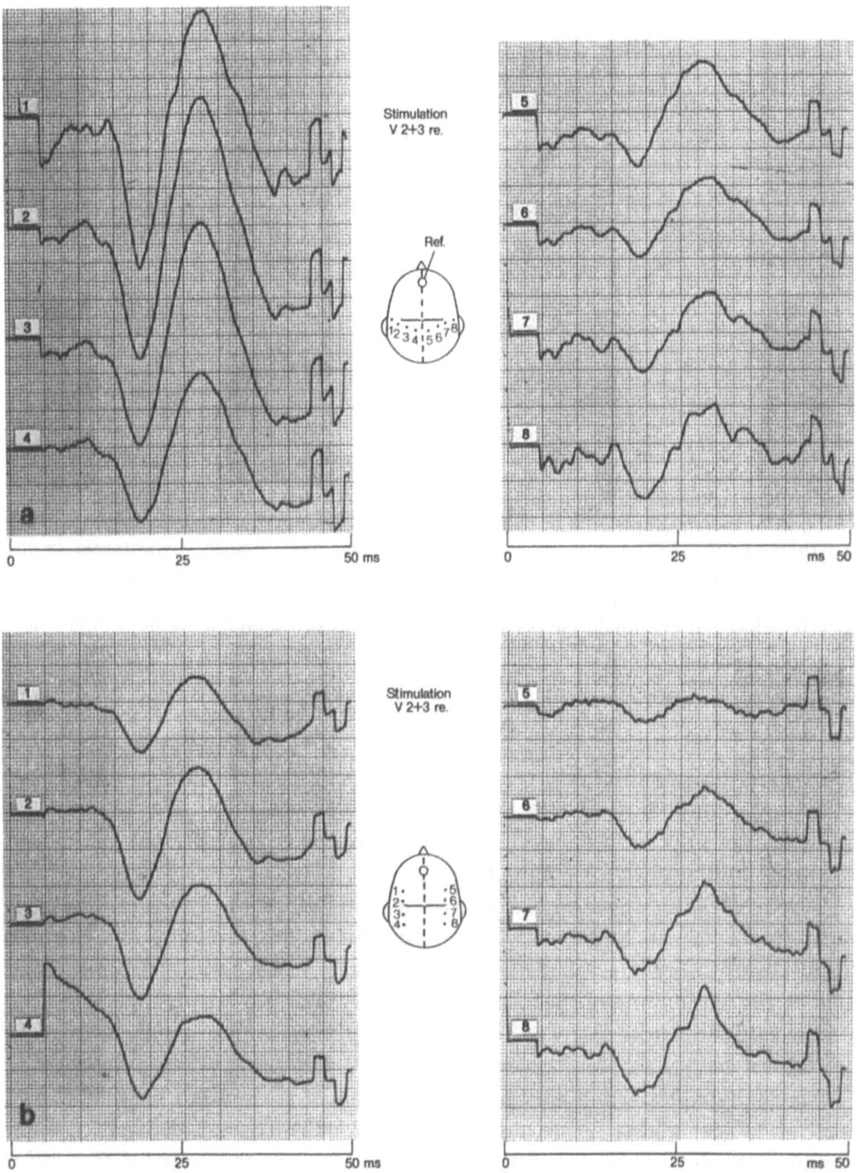

Fig. 2a,b. Scalp distribution of the cortical response after trigeminal stimulation (simultaneous, unilateral stimulation of the upper and lower lips; registration against a frontomedian reference). **a** Transverse series registration over the ipsilateral and contralateral hemisphere. Stimulus responses show a maximum over the temporal positions of the hemisphere contralateral to the stimulation side, and increasingly decrease in amplitude towards the sagittal line as well as over the ipsilateral hemisphere. **b** Longitudinal series registration 9 cm lateral to the sagittal line over the ipsilateral and contralateral hemisphere. Maximum of the potential amplitude over the contralateral hemisphere in the crown of the head/auricular line

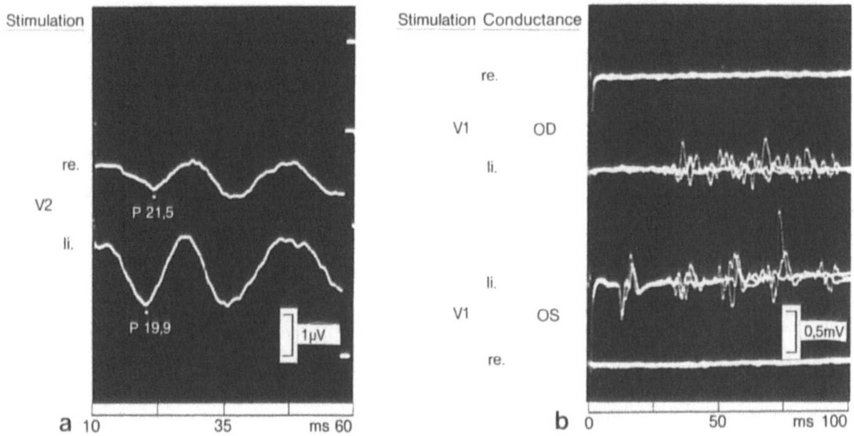

Fig. 3 a,b. Trigeminal SEP in incomplete peripheral trigeminal lesion. A 30-year-old man with partly resected chordoma in the middle cranial fossa on the right with postoperative anesthesia in the region innervated by the first trigeminal branch, hypesthesia in the region innervated by the second trigeminal branch on the right. After upper lip stimulation on the left side (**a**), the cortical response is normal. It is moderately lowered and slightly delayed after stimulation on the right side (**b**). Corresponding to the complete loss of sensitivity in the skin area innervated by the first trigeminal branch, the ipsilateral and contralateral blink reflex is lost after stimulation of the right supraorbital nerve and is normal after stimulation on the contralateral side

more than 22.2 ms or a difference from one side to the other of more than 1.93 ms is pathological (Table 1). With regard to the amplitude, a unilateral reduction in amplitude of more than 40% can be regarded as an indication of partial interruption of conduction.

In Wallenberg's syndrome, our own investigations of the trigeminal SEP showed a delay in latency and/or reduction in amplitude of the cortical stimulus responses after trigeminal stimulation in the mouth in more than half of the patients (Fig. 4). This finding is somewhat surprising, since the lemniscal conduction pathway described at the beginning is not affected in the lateral medulla oblongata syndrome of Wallenberg. Accordingly, there is merely hypalgesia and thermal hypesthesia of the ipsilateral half of the face in typical cases. With a subtle test of sensitivity, however, a slight impairment also of the epicritical functions (e.g., in the form of two-point discrimination at the lips differing from one side to the other) can sometimes be detected. This must be regarded as an indication of the not uncommon spreading of the infarct into the pontine tegmentum. It is also indicated by the blink reflex findings with lowering and not only delay of the late reflex components after stimulation on the damaged side, but also a delayed and lowered early reflex component which often occurs in such patients and which is known to indicate the presence of a pontine lesion [10]. The pathological trigeminal SEP findings which are often demonstrated in patients with Wallenberg's syndrome are thus to be rated as an indication of

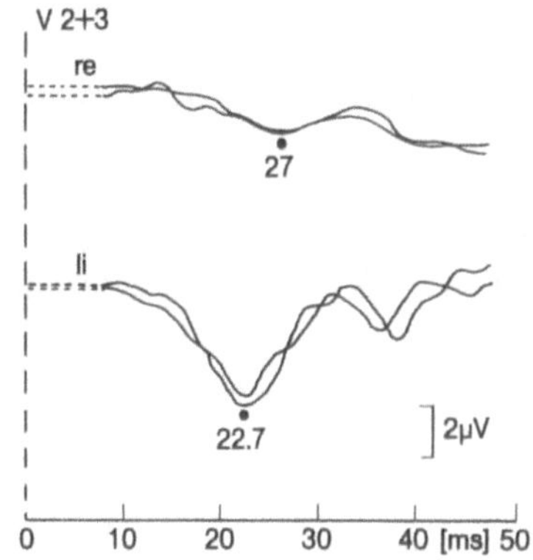

Fig. 4. Trigeminal SEP in the Wallenberg syndrome. A 58-year-old man with lateral medulla oblongata syndrome with pronounced hypalgesia and thermal hypesthesia in the area innervated by the right trigeminal nerve and discrete disorder of epicritical sensitivity. The cortical response after stimulation of the upper and lower lip on the right side is reduced and pathologically delayed. *re*, right; *li*, left

Table 1. Normal values of the primary cortical response after trigeminal stimulation of the lips

A Mean latencies of N13, P19, and N27, and mean amplitudes of the primary cortical response (N13/P19)

	Latencies (ms)			Amplitude (μV)
	N13	P19	N27	N13/P19
Mean value	12.5	18.5	26.9	2.6
SD	0.87	1.51	2.23	1.0
Upper limit (mean + 2.5 SD)	14.7	22.3	32.5	

B Differences between the two sides for the latencies and the amplitude N13/P19 ($n = 82$ for N13 and P19, $n = 80$ for N27, $n = 41$ for the differences between sides). [From 8]

	Difference of the latencies from one side to the other (ms)			Difference of the amplitude (μV) from one side to the other
	N13	P19	N27	N13/P19
Mean value	0.45	0.55	0.72	0.51
Standard deviation	0.35	0.55	0.63	0.54
Upper limit (mean + 2.5 SD)	1.34	1.93	2.29	1.86

spreading of the infarct into the pontine tegmentum, passing beyond the classical area, and not as an indication that the trigeminal SEP are conducted via the spinal pathway and spinal nucleus of the trigeminal nerve.

Trigeminal SEP Findings in Multiple Sclerosis

Our own investigations in 46 patients with confirmed multiple sclerosis showed a unilateral or bilateral pathological increase of the absolute latency of P19, a pathological difference of P19 from one side to the other, or a unilateral failure of the stimulus response in 61%. This method thus shows a diagnostic precision comparable to that of the registration of acoustic evoked brain-stem potentials [7]. The fact that half of the patients with pathological findings did not show any disturbance of sensitivity in the face is of particular interest. In the patient population reported by Eisen et al. [5], 41.4% of the patients with probable or confirmed multiple sclerosis

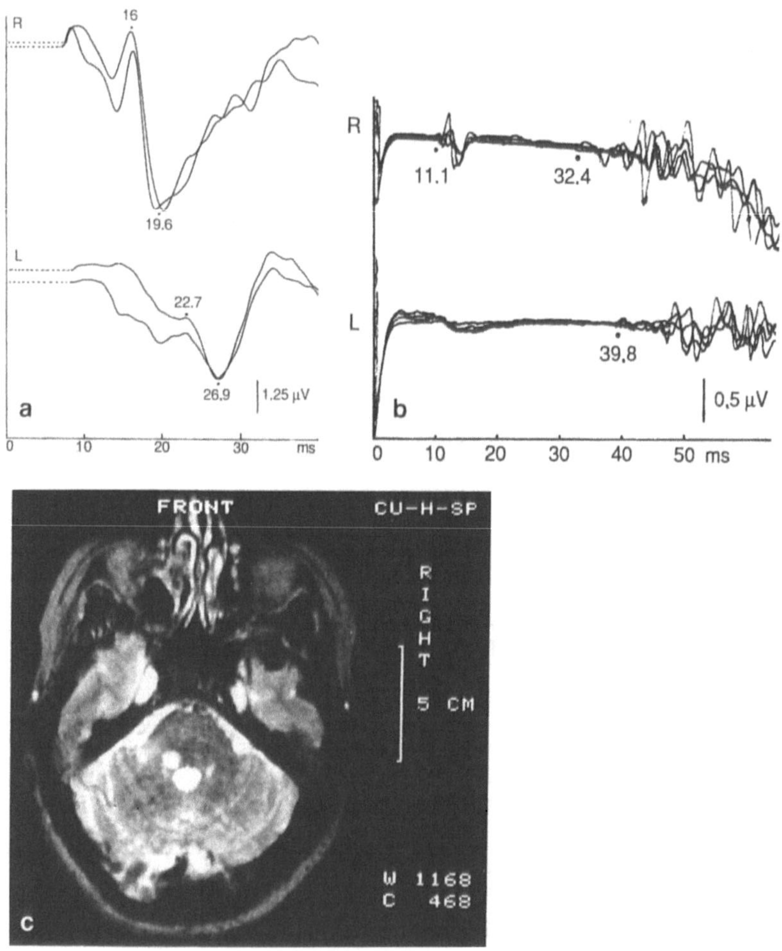

Fig. 5a–c. Trigeminal SEP in multiple sclerosis. **a** Delay of the cortical response (P19) after trigeminal stimulation on the left; difference between the two sides: 7.3 ms. **b** The registration of the blink reflex shows a loss of the early reflex and a delay of the late reflex as an indication of a pontine as well as an additional medullary lesion after stimulation and registration on the left side. **c** In MRI, a focus in the left pons is shown as the correlate of the P19 delay and the loss of the early reflex response. The focus additionally present in the medulla oblongata owing to the delayed late blink reflex component was not detectable

without clinical indications of an involvement of the trigeminal pathways also had pathological trigeminal SEP. In clinically purely spinal forms of multiple sclerosis, such a finding is especially helpful, since it enables identification of a supraspinal focus. In lesions preferentially located in the brain stem, the demyelinating character of the disease can be documented by a

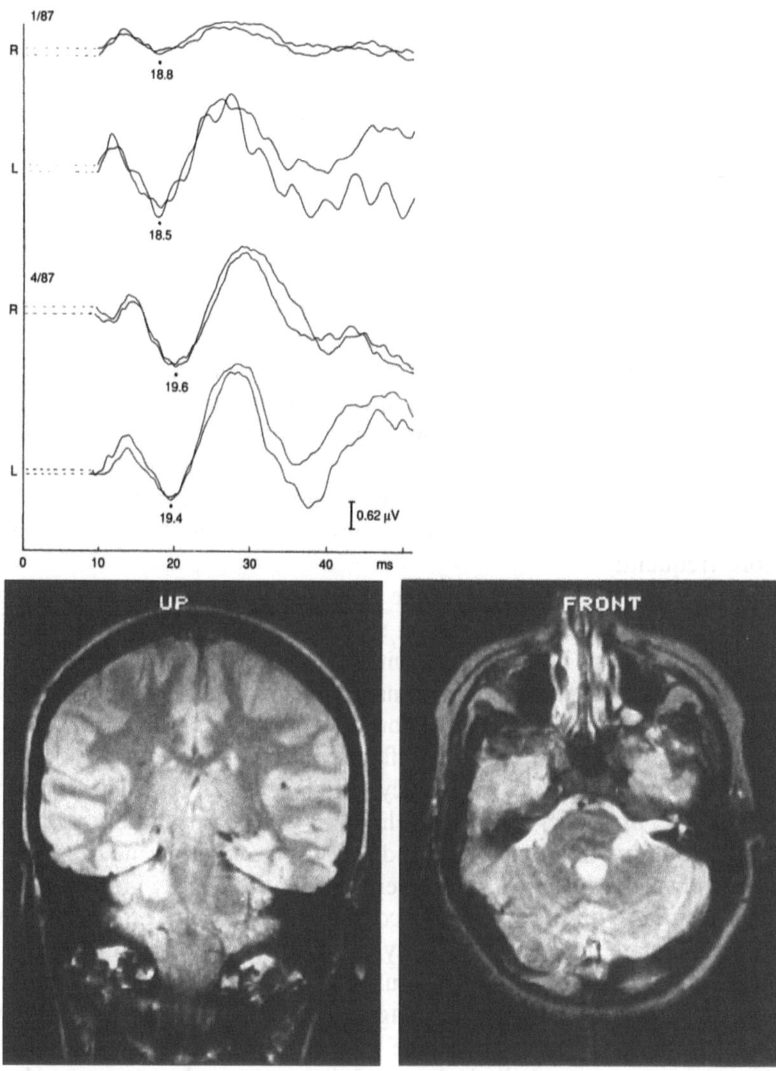

Fig. 6. Trigeminal SEP in multiple sclerosis with partial blockade of conduction on its own in the region of the right pontine focus, with significant reduction in amplitude of the cortical response after stimulation of the lower lip on the right side (1/87). The follow-up registration 3 months later (4/87) shows normalization of the findings. MRI shows a hyperintensive lesion of the lateral pons on the right. Coronal (*left*) and horizontal (*right*) sections. [From 10]

more pronounced increase in latency, which is not shown by CSF investigation or by MR imaging.

In the example in Fig. 5, the investigation of the trigeminal SEP shows a pronounced increase in latency of P19 after stimulation of the left side. This finding corresponds well with a left pontine focus of demyelination visible on MRI. The early reflex component of the blink reflex cannot be definitively registered at the stimulation and registration on the left, which also conforms well with the lesion in the left pons. However, a delayed blink reflex can be found in addition as indication of a further focus in the lateral medulla oblongata. A focus with this localization cannot be detected with the method of the trigeminal SEP, and in this case was also not detectable by means of MRI; this applies more frequently in particular in patients with small brain-stem foci.

Delays in latency of P19 are a quite typical, but not obligatory, finding in patients with multiple scleosis. In recent foci of demyelination with isolated conduction block, mere reductions in amplitude up to potential loss occur, and complete recovery may take place in such a case.

In the example in Fig. 6, the upper curve pair shows a significant lowering of amplitude after stimulation of the lower lip on the right side; this corresponds to a hyperintense focus in the right pons. The follow-up investigation carried out 3 months later (lower curve pair) shows normalization of findings as a manifestation of optimal remyelination in this patient. However, defect healing with permanent disorders of impulse conduction results more frequently.

In the investigation of multiple sclerosis patients, use of imaging techniques alone is often considered sufficient, and diagnosis of neurophysiological function is dispensed with. However, foci visible on CT and MRI are as such completely nonspecific and occur in the most diverse neurological disorders; in the region of the brain stem in particular it is not possible to distinguish between inflammatory and ischemic lesions. In contrast, a pronounced delay in latency visible in the trigeminal SEP practically proves the presence of a demyelinating lesion; i.e., there are not only localization, but also etiological indications. In addition, neurophysiological diagnostic procedures, which are very efficient especially in the region of the brain-stem, are often sufficient in combination with the CSF investigation to attain adequate diagnostic certainty in multiple sclerosis patients with whom there are consistent clinical features. For these patients, elaborate and expensive diagnostic tests with imaging techniques can be dispensed with.

Altogether, registrations of the trigeminal SEP are suitable to demonstrate pontomesencephalic lesions as long as the main sensory nucleus and/or the specific trigeminal pathways are affected. In neoplastic and ischemic lesions, reductions in amplitude, and in multiple sclerosis, delays in latency of the primary cortical response predominate. Apart from aiding localization, this method also provides etiological information and can be used for objective establishment of a sensibility disorder in the face.

References

1. Altenmüller E, Cornelius CP, Buettner UW (1990) Somatosensory evoked potentials following tongue stimulation in normal subjects and patients with lesions of the afferent trigeminal system. Electroencephalogr Clin Neurophysiol 77:403–415
2. Bennett MH, Jannetta RJ (1980) Trigeminal evoked potentials in humans. Electroencephalogr Clin Neurophysiol 48:517–526
3. Darian-Smith I (1973) The trigeminal system. In: Autrum H, Jung R, Loevenstein WR (eds) Somatosensory system. Springer, Berlin Heidelberg New York, pp 271–314 (Handbook of sensory physiology, vol 2)
4. Drechsler F (1980) Short and long latency cortical potentials following trigeminal nerve stimulation in man. In: Barber C (ed) Evoked potentials. MTP Press, Lancaster, pp 415–422
5. Eisen A, Paty D, Purves S, Hoirch M (1981) Occult fifth nerve dysfunction in multiple sclerosis. Can J Neurol Sci 8:221–225
6. Ishiko N, Hanamori T, Murayama N (1980) Spatial distribution of somatosensory responses evoked by tapping the tongue and finger in man. Electroencephalogr Clin Neurophysiol 50:1–10
7. Robinson K, Rudge P (1977) Abnormalities of the auditory evoked potentials in patients with multiple sclerosis. Brain 100:19–40
8. Stöhr M, Petruch F (1979) Somatosensory evoked potentials following stimulation of the trigeminal nerve in man. J Neurol 220:95–98
9. Stöhr M, Petruch F, Scheglmann K (1981) Somatosensory evoked potentials following trigeminal nerve stimulation in trigeminal neuralgia. Ann Neurol 9:63–66
10. Stöhr M, Dichgans J, Diener HC, Buettner UW (1989) Evozierte Potentiale. SEP-VEP-AEP-EKP-MEP, 2nd edn. Springer, Berlin Heidelberg New York

Diagnostic Value of Trigeminal SEP and the Jaw-Opening Reflex in Brain-Stem Lesions

E. Altenmüller, U.W. Buettner, and J. Dichgans

Neurologische Klinik, Kliniken Schnarrenberg, Eberhard Karls-Universität, Hoppe-Seyler-Straße 3, 72076 Tübingen, Germany

Introduction

Trigeminal somatosensory evoked potentials (SEP) following electrical stimulation of the lips were introduced into electrophysiological routine examination by Stöhr and Petruch [8], who showed that a prominent positive wave over the contralateral somatosensory face area at a latency of 19 ms (P19) could reliably be recorded in all normal subjects. This wave was pathologically delayed in 60% of patients who had clinically definite multiple sclerosis (MS), a finding demonstrating the usefulness of trigeminal SEP in the diagnosis of demyelinating diseases [3]. Although anecdotal reports on the sensitivity of trigeminal SEP in brain-stem disorders exist [3, 7], no systematic evaluation has been undertaken up to now. This presentation reports on an improved technique of stimulation [1, 2] and compares the diagnostic value of trigeminal cortical SEP and of the jaw-opening reflex in patients with disturbances of facial sensibility due to brain-stem lesions of different causes.

Patients and Methods

Twenty patients aged 18–65 years (mean 43 years) with brain-stem lesions were investigated. All of them had clinical symptoms of uni- or bilateral trigeminal dysfunction, i.e., hypo- or anesthesia and hypo- or analgesia of the face. Additional trigeminal signs included unilateral paresis of the masseter and pterygoid muscles in four patients. In all patients CT or MRI scans showed a lesion in the brain stem. Patients were subdivided into three diagnostic categories according to their clinical history, neurological examination, and laboratory and neuroradiological findings. Seven patients had clinically definite MS according to the criteria of McAlpine et al. [6]. In this group, additional symptoms included oculomotor abnormalities and motor deficits, sensory deficits, and reduced visual acuity due to optic neuritis. Six patients had isolated brain-stem lesions due to a unifocal inflammatory

L.R. Caplan and H.C. Hopf (Eds.)
Brain-Stem Localization and Function

process. Additional symptoms varied with the extension of the process and included oculomotor symptoms, facial paresis, glossopharyngeal paresis, and disturbance of audition. Seven patients had ischemic lesions of the medulla oblongata and pons with typical sudden onset of symptoms. All of these patients had additional signs such as homolateral hemiataxia, glossopharyngeal paresis, and contralateral dissociated disturbance of sensibility.

Stimulation was performed with a new method developed recently for stimulation of the lingual nerve [1]: modified EEG earclip electrodes (5 mm diameter) were placed on either side of the upper lip (infraorbital nerve), the lower lip (inferior alveolar nerve), or the tip of the tongue (lingual nerve). Stimuli consisted of 0.1 ms square wave pulses delivered at a rate of 3/s. Intensity was set at four times the sensory threshold, ranging from 2 to 5 mA; 512 or 1024 stimulus responses were averaged. Stimuli were not painful, but they caused a visible local contraction of the muscle above which the electrode was attached and not of neighboring muscles.

Subcutaneous platinum needle electrodes were placed above the contralateral face area of the somatosensory cortex, between $C_{3/4}$ and $T_{3/4}$ according to the 10/20 system. To record the jaw-opening reflex following stimulation of the tongue, electrodes were placed above the pterygoid muscles at the ipsilateral tragus. All electrodes were referred to F_z.

Since in normal subjects the prominent and most reliable component of trigeminal SEP consisted of a positivity at about 19 ms latency [1, 8], further analysis was restricted to this positive wave of cortical origin. Findings were considered pathological when responses were absent, or when the latencies were more than three standard deviations greater than mean values obtained in 20 normal subjects. According to this criterion, the normal limits of latencies for SEP following stimulation of the upper and lower lips were 22.4 ms, after stimulation of the tongue 26 ms. The upper limit of latency for the jaw-opening reflex in normal subjects was 30 ms. Due to the large intraindividual variability of amplitudes when the left or right side of the face is stimulated unilateral amplitude reduction of potentials was not accepted as a criterion for pathology.

Results

Figure 1a shows typical responses in a normal subject and in a patient with left-sided hypoesthesia of the face and the tongue due to a small, left-sided pontomedullary inflammatory focus shown on MRI (Fig. 1b). Neurological examination and CSF revealed no further abnormalities; brain-stem auditory evoked responses (BAER) and tibial SEP were normal. The responses following stimulation of the upper and the lower lip are well preserved, whereas the lingual SEP and the jaw-opening reflex are abolished. It should be noted that in both the normal subject and the patient, stimulation of the lips yields large stimulus artifacts, whereas stimulation of the tongue

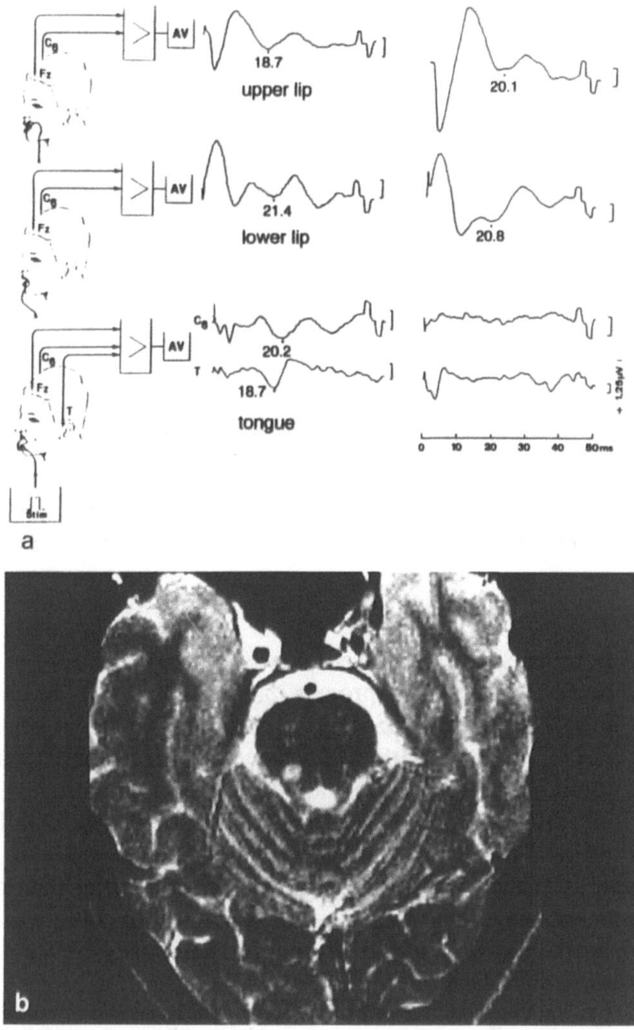

Fig. 1. **a** Trigeminal SEP following stimulation of the left upper lip, lower lip, and tongue in a nomal subject (*left*) and in the patient EH (*right*). Lingual SEP are abolished in the patient (*third trace*, right). Equally, the jaw-opening reflex is not elicited (*lowest trace*). Note the large stimulus artifacts caused by stimulation of the upper or the lower lip. **b** T_2-weighted MRI image (TR = 2.1 s, TE 90 ms) of patient EH revealing a small, left-sided pontomedullary inflammatory focus

produces no stimulus artifact, thus allowing the assessment of earlier components such as N13.

Figure 2 summarizes the results of cortical SEP and the jaw-opening reflex in the three groups of patients. Fifteen of the 20 patients showed abnormalities of at least one of the four parameters tested. In general, the most common pathological finding was the abolition of responses. Latencies

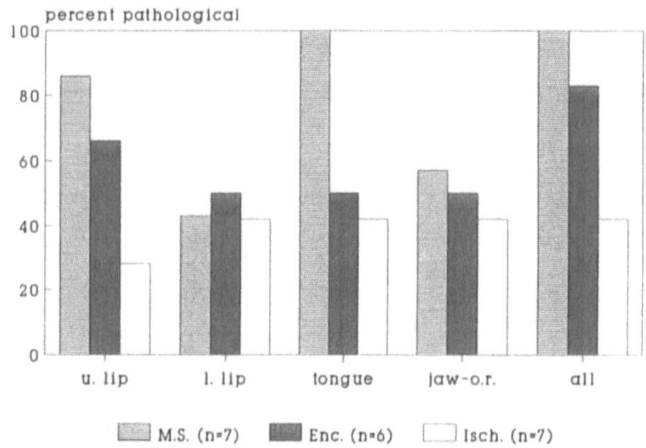

Fig. 2. Incidence of pathological findings in the different diagnostic categories (*MS*, multiple sclerosis; *Enc*, isolated brain-stem encephalitis; *Isch*, ischemic lesion of the brain stem). The four stimulation methods are compared (*u. lip*, stimulation of the upper lip; *l. lip*, stimulation of the lower lip; *jaw-o.r.*, jaw-opening reflex). *All* denotes the percentage of pathological findings in at least one of the parameters tested

of potentials were increased in two patients with MS, in one patient with a unifocal inflammatory process, and in one with an ischemic brain-stem lesion. The highest incidence of pathological findings was observed in the group of patients with MS when the lingual SEP were recorded (7/7). Potentials following stimulation of the lower lip were less sensitive in the MS group (three of seven patients). In patients with unifocal inflammatory processes, the sensitivities of SEP and the jaw-opening reflex were not as high as in the MS group, but they were still better than in the patients with ischemic brain-stem lesions. Although all of the latter patients suffered from an ipsilateral diminution of epicritic sensibility of the face, the responses were pathological in only three.

Comparing the overall sensitivity of trigeminal cortical SEP with the jaw-opening reflex, there was no difference with the exception of the MS group: all seven MS patients had a loss or delay of lingual SEP, but only four of them showed a pathological jaw-opening reflex.

Discussion

The diagnostic value of trigeminal somatosensory evoked potentials and of the jaw-opening reflex in brain-stem lesions is still a matter of debate. In contrast to far-field BAER, trigeminal SEP and the jaw-opening reflex have not gained general acceptance as diagnostic tools. The most prominent obstacle to using trigeminal SEP as an indicator of brain-stem dysfunction is technical: electrical stimulation in facial regions may lead to large stimulus artifacts, which mask short-latency evoked potentials arising from the somatosensory trigeminal pathway in the periphery and the brain stem. Furthermore, it has recently been argued that the "late" scalp responses of about 20 ms latency, which are believed to arise from the somatosensory cortex, are contaminated with muscular artifacts and are therefore of little diagnostic value [4].

The present study demonstrates that somatosensory trigeminal SEP of cortical origin and the jaw-opening reflex are useful in the objective assessment of dysfunction of the afferent somatosensory pathway in patients with brain-stem lesions. Cortical components can be distinguished from reflex activity arising from masticatory muscles by synchronous monitoring of the muscle activity at the tragus. The stimulation and recording procedures can easily be performed in any routine laboratory. Concerning the problem of electrical stimulus artifact, the best results are obtained by stimulation of the lingual nerve at the tip of the tongue. Since this method of stimulation yields no baseline drift, assessment of good-quality early far-field components is possible. As with the "brain-stem" far-field potentials P13–P14 elicited by stimulation of the median or ulnar nerve [5], the use of a noncephalic reference in lingual SEP might disclose components generated in the trigeminal root or the trigeminal nuclei, thus improving the topodiagnostic value of lingual SEP considerably.

Owing to the small number of patients investigated to date, statistical evaluation concerning the differential diagnostic value of trigeminal SEP was not performed. As a general tendency, there is a better electroclinical correlation in MS patients and patients with inflammatory processes, whereas in vascular brain-stem lesions the sensitivities of trigeminal SEP and the jaw-opening reflex seem to be relatively poor. Similar results have been published by Stöhr [7], who found that only about 50% of his patients with Wallenberg's syndrome had pathological responses when stimulated on the upper and the lower lip simultaneously. Unfortunately, no clinical correlations are reported in that study.

A surprising and hitherto unexplained result is the relatively low incidence (50%) of abnormal jaw-opening reflexes. Since this polysynaptic reflex relies on a pathway which includes, in addition to the structures mediating the somatosensory potential, the motor nucleus and probably the mesencephalic trigeminal nucleus, we expected a higher sensitivity in brain-stem lesions compared with the trigeminal somatosensory evoked potential. To further establish trigeminal SEP and trigeminal reflexes as topodiagnostic tools in patients with brain-stem lesions, more clinical correlation studies combining the measurement of brain-stem far-field potentials and the precise localization of lesions by MRI examination are needed.

Acknowledgement. We are grateful to Dr. T. Nägele for performing the MRI and Mrs. C. Winkler and Mrs. Heidi Uhl for their expert technical assistance in performing the SEP examinations.

References

1. Altenmüller E, Cornelius CP, Buettner UW (1990) Somatosensory potentials following tongue stimulation in normal subjects and in patients with lesions of the afferent trigeminal system. Electroencephalogr Clin Neurophysiol 77:403–415
2. Altenmüller E, Roehlich M, Winkler C, Buettner UW (1991) Wozu Trigeminus SEP? Diagnostische Wertigkeit bei Erkrankungen der somatosensiblen trigeminalen Afferenz. Z EEG EMG 21:99
3. Buettner UW, Petruch F, Scheglmann K, Stöhr M (1982) Diagnostic significance of cortical somatosensory evoked potentials following trigeminal nerve stimulation. In: Courjon J, Maugiere F, Revol M (eds) Clinical applications of evoked potentials in neurology. Raven, New York, pp 339–345
4. Leandri M, Parodi CI, Zattoni J, Favale E (1987) Subcortical and cortical responses following infraorbital nerve stimulation in man. Electroencephalogr Clin Neurophysiol 66:253–262
5. Maugiere F (1987) Short-latency somatosensory evoked potentials to upper limb stimulation in lesions of brainstem, thalamus and cortex. In: Halliday AM (ed) The London symposia. Elsevier, Amsterdam, pp 302–309
6. McAlpine D, Lumsden CE, Acheson ED (1972) Multiple sclerosis: a reappraisal. Churchill Livingstone, Edinburgh
7. Stöhr M (1989) Somatosensible Reizantworten bei Erkrankungen des peripheren und Zentralen Nervensystems. In Stöhr M, Dichgans J, Diener HC, Buettner UW (eds) Evozierte Potentiale. Springer, Berlin Heidelberg New York, pp 129–249
8. Stöhr M, Petruch F (1979) Somatosensory potentials following stimulation of the trigeminal nerve in man. J Neurol 220:95–98

The Masseter Reflex in the Topodiagnosis of Brain-Stem Lesions

R. Dengler

Neurologische Klinik, Medizinische Hochschule Hannover, Konstanty-Gutschow-Straße 8, 30625 Hannover, Germany

Introduction

The trigeminal masseter reflex (MassR) is a true monosynaptic brain-stem reflex. It plays an important role in routine neurological examination and can be of great value, e.g., in the differentiation of spasticity of supraspinal and of spinal origin. Basically, the MassR is a unilateral reflex, the responses of the two sides being independent of each other [3]. Clinically, however, we observe only the jaw movement, and diagnostically important differences between the responses of the two masseter muscles can hardly be evaluated. Electromyographic analysis of the MassR provides a means of assessing the unilateral reflex responses selectively, especially with regard to quantitative parameters such as latency and amplitude. This makes the electromyographic recording of the MassR, whose central pathways run through pontine and mesencephalic areas, a valuable diagnostic tool in brain-stem disorders. The following contribution is an attempt to illustrate the role of the electromyographically recorded MassR in the topodiagnosis of brain-stem lesions.

Anatomy and Physiology of the MassR

A scheme of the reflex pathways is shown in Fig. 1. The reflex is elicited by stretch of structurally typical muscle spindles [2] in the masseter muscle. The afferent limb is formed by the third branch of the trigeminal nerve, although there is still some debate on whether the Ia fibers enter the pons through the motor [9] or through the sensory root [10] of the nerve. In the pons the axons turn upward, following the mesencephalic tract of the trigeminal nerve to the cell bodies of their mother cells (large unipolar ganglion cells) in the mesencephalic nucleus [14]. The other axon collateral of the unipolar ganglion cells turns back to the pons to make monosynaptic contact [14] with motoneurons in the masticatory nucleus. The efferent reflex limb is formed by the masticatory nerve, which travels adjacent to the third branch of the trigeminal nerve to the muscles of the chewing apparatus. Thus,

L.R. Caplan and H.C. Hopf (Eds.)
Brain-Stem Localization and Function

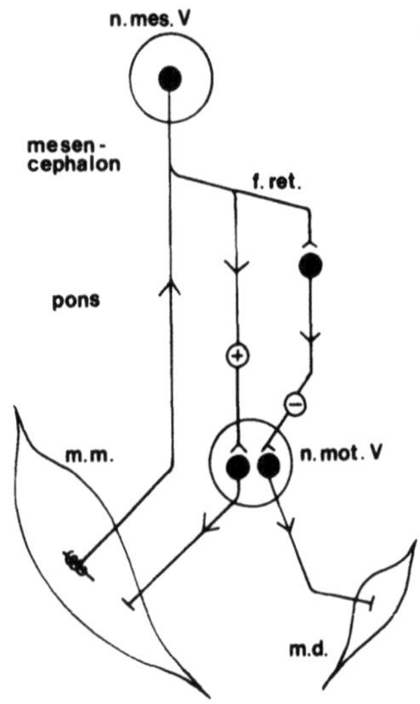

Fig. 1. Pathway of the masseter reflex. One axon collateral (Ia fiber, primary spindle afferent) of the pseudounipolar ganglion cell in the mesencephalic trigeminal nucleus (*n.mes.V*) innervates a muscle spindle in the masseter muscle (*m.m.*); the other collateral makes monosynaptic contact with motoneurons in the trigeminal motor nucleus (*n.mot.V*) in the pons. A parallel oligosynaptic inhibitory connection to motoneurons of the digastric muscle (*m.d.*) through the reticular formation (*f.ret.*) is questionable [2]

the reflex pathways involve the third branch of the trigeminal nerve and pontomesencepalic brain-stem areas.

The MassR is monosynaptic [14] and has a correspondingly short and stable latency (6–9 ms). It is not clear whether the positive reflex response in the masseter is associated with reciprocal inhibition of the ipsilateral jaw-opening muscles [3], i.e., of the digastric muscle, similar to the spinal reflex connectivity. Although the reflex is unilateral [9], there may be connections to the other side of the brain-stem mediating facilitation to the motoneurons in the contralateral masticatory nucleus.

The Normal MassR: Methods and Findings

In electromyographic analysis, the reflex is elecited by a tap to the chin, using a hammer with a trigger device. The reflex is tested with the masseter muscles in a relaxed state, i.e., with the mouth slightly opened. If there is no detectable reflex, subjects are asked to innervate the muscles. The responses are usually recorded by means of surface techniques with the active electrode placed over the belly of the masseter muscle and the reference over a virtually inactive area, e.g., the side of the nose. The EMG signals are amplified using filters appropriate for surface recording (e.g., bandwith from 20 Hz to 3000 Hz). In our lab, we analyze the reflex at least five times and routinely record a short period of masseter EMG during maximal voluntary contraction. The latter is used to check for side-to-side differences which may be caused artificially by electrode problems or biologically by decreased motor unit recruitment, i.e., peripheral denervation. A normal MassR is illustrated in Fig. 2, trace 1.

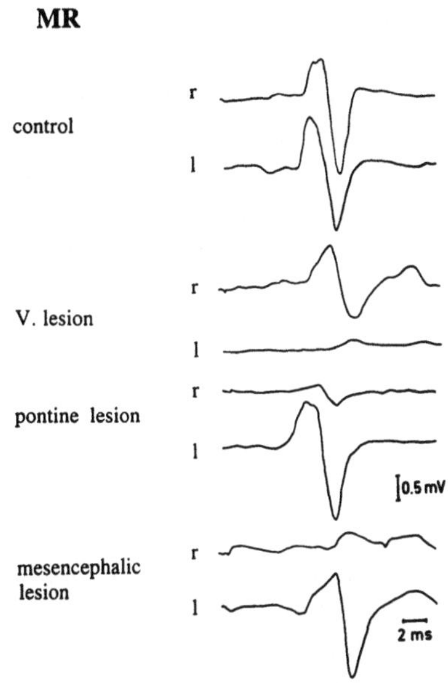

Fig. 2. MassR of a control and of patients with a left peripheral trigeminal (*V*), a right pontine, and a right mesencephalic lesion. All patients exhibited a more than 50% decrease of the reflex amplitude on the affected side and a latency side-to-side difference of more than 1 ms

The latency of the onset of the electromyographic response is essentially influenced by the delay of the trigger device. In our lab it is accepted as normal up to 9.0 ms, in agreement with other authors [15]. Side-to-side differences must not exceed 1 ms. Amplitudes are very variable and depend on factors such as stimulus strength [8] and degree of muscle pre-innervation. Therefore, we do not use absolute amplitudes for diagnostic purposes. Typical values range from 0.5 mV to 3 mV. Amplitude side-to-side differences can be considerable. We empirically regard differences greater than 50% as abnormal.

The MassR in Pathological Conditions

Trigeminal Lesions

Lesions of the trigeminal nerve, especially of its third branch or the motor nerve, can be associated with either prolongation of the reflex latency or reduction of the amplitude [1, 7]. The MassR abnormality illustrated in Fig. 2, trace 2, was caused by an osteodestructive process in the base of the middle cerebral fossa and partially in the temporal bone involving the second and third branch of the nerve. On presentation the patient suffered only from pain in the area of these two trigeminal branches, which developed several weeks after she had experienced a diagnostically misleading transient facial weakness on the same side. EMG of the left masseter muscle revealed signs of mild acute denervation.

Differentiation between peripheral and central lesions, i.e., pontine or mesencephalic, is mostly possible on the basis of typical clinical findings. Further arguments in favor of a peripheral disorder are signs of denervation in masseter needle EMG if the masticatory nerve is also affected. Although the MassR may frequently be abnormal in peripheral trigeminal disease and may therefore be used as an instrument in trigeminal nerve diagnosis [1, 7], it is clearly more valuable in the assessment of brain-stem disorders.

Pontine Lesions

The MassR abnormality illustrated in Fig. 2, trace 3, was caused by a unilateral pontine vascular malformation located mainly in the tegmentum. Clinically, the patient showed a pure sensory trigeminal deficit, without further brain-stem signs and without indication of denervation in the masseter EMG. Pontine lesions may cause MassR changes by damaging either spindle afferents on their way to the mesencephalon through the upper tegmentum pontis, or efferent reflex structures involving the motor nucleus and the intrapontine portion of the motor axons. The latter would, in most cases, be associated with pathological findings in needle EMG, especially with fibrillation potentials or positive sharp waves. The lower mesencephalic tract, the main sensory nucleus, and the upper spinal tract of the trigeminal nerve are closely adjacent anatomical structures in the tegmentum pontis. Since the latter two trigeminal relay stations are involved in the pathways of the early and late BR component, respectively, pontine lesions of trigeminal afferents will frequently be associated with changes in the ipsilateral blink reflex [11]. These may consist of abnormalities of the R1 component or both R1 and R2, as illustrated in Fig. 3. In this case, both the R1 component and the bilateral R2 were delayed following stimulation on the left side. The MassR was almost absent on the left side, but there was a good response on the right. This patient had definite multiple sclerosis and showed a small area of increased signal intensity on T_2-weighted MRI images in the right tegmentum pontis. Multiple sclerosis is probably the most frequent cause of the combination of MassR and Blink reflex abnormalities.

Mesencephalic Lesions

Lesions involving the mesencephalic trigeminal tract and/or nucleus will mostly alter the MassR. Depending on the type of lesion, i.e., demyelinating or primary axono-degenerative, one can find latency prolongations, amplitude reductions, a combination of both, and even loss of the responses. The patient whose MassR (Fig. 2, trace 4) exhibited a marked reduction of the amplitude on the right had a CT-proven acute right-sided midbrain hemorrhage. Typically, blink reflexes and masseter needle EMG provided completely normal results, as usual in selective midbrain processes.

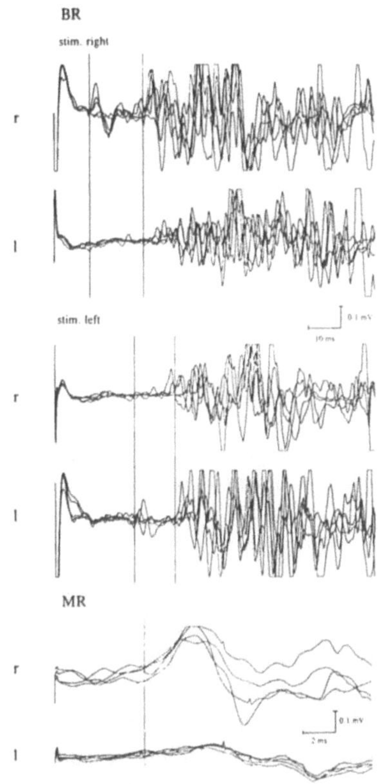

Fig. 3. MassR (*MR*) and blink reflex (*BR*) of a patient with multiple sclerosis. Both R1 and R2 of the blink reflex are delayed following left-side stimulation. The left MassR is almost absent. The patient revealed an area of increased signal intensity in the left tegmentum pontis on MRI

Since the mesencephalic tract and nucleus of the trigeminal nerve have a close anatomical relationship to the oculomotor nuclei and the medial longitudinal fascicle (MLF) they are frequently involved in lesions of these structures. The MassR can therefore be helpful in deciding whether oculomotor abnormalities are caused by a peripheral or a brain stem lesion. Recently, Hopf and Gutman [5] were able to confirm that diabetic third nerve palsy with and without pupillary sparing is more commonly caused by ischemia of the nerve within the mesencephalon than by infarction in the peripheral course of the nerve. Ten of their 11 patients had MassR changes ipsilateral to the side of the oculomotor deficit which could not be explained by a peripheral trigeminal deficit and thus indicated a midbrain lesion. A control group of 13 patients with Weber's syndrome, another type of vascular midbrain lesion, all had ipsilateral MassR abnormalities. MassR abnormalities have also been described in midbrain tumors [4, 11].

The MassR with an almost absent response on the left side, illustrated in Fig. 4, was recorded in a 35-year-old man who presented clinically with the acute onset of internuclear opthalmoplegia on the same side. The remainder of the neurological examination and previous history were unremarkable. Blink reflexes were normal with regard to both early and late components.

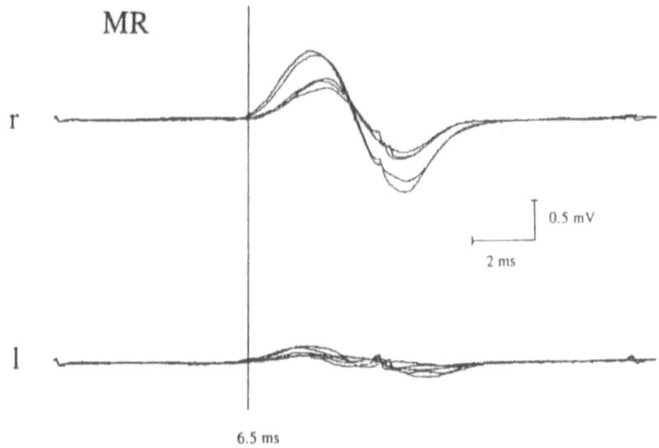

Fig. 4. MassR (*MR*) of a patient with internuclear ophthalmoplegia. The left response exhibits an amplitude clearly less than 50% of the right

A CT scan and MRI showed no abnormalities including the brain stem. Our diagnosis was "possible multiple sclerosis", although the cerebrospinal fluid was also normal. On the basis of the MassR findings, we concluded that the patient had a small lesion in the left mesencephalic tegmentum involving the MLF and the mesencephalic trigeminal tract and/or nucleus. This case confirms the findings of other authors [6], that careful analysis of both blink reflex and MassR in patients with internuclear ophthalmoplegia can be helpful in separating lesions of the MLF at the pontine level from those at the mesencephalic level.

Suprasegmental Lesions, Infectious Tetanus

Similar to the stretch reflexes of the extremities, the MassR can be exaggerated in lesions of the corticospinal or corticobulbar tract. There are, however, no available systematic EMG studies of changes of the MassR in suprasegmental lesions. Such lesions may induce side-to-side differences of the reflex amplitudes but will not result in prolongation of reflex latencies or unilateral reduction of amplitude. Therefore, confusion of MassR abnormalities due to suprasegmental and brain-stem disorders appears very improbable.

A prominent and early symptom of infectious tetanus is masseter spasms, so-called trismus. The MassR is not exaggerated in this situation, indicating that interneurons not involved in the reflex pathway are affected. The post-reflex silent period, however, is shortened or even lacking [11]. The latter finding can be helpful in the early diagnosis of tetanus.

Conclusions

The MassR is unique in that it is the only electromyographically recordable cranial reflex with a pathway involving both pontine and midbrain areas. The recording techniques are simple and not invasive. Therefore, the MassR is a valuable tool in the topodignosis of brain-stem disorders, especially if applied in combination with other neurophysiological tests such as the blink reflex, needle EMG of facial and chewing muscles, auditory evoked brainstem potentials, and electro-oculography.

References

1. Dengler R, Struppler A (1981) Neurophysiological diagnosis of trigeminal nerve function. In: Samii M, Janetta PJ (eds) The cranial nerves. Springer, Berlin Heidelberg New York, pp 302–311
2. Freimann R (1954) Untersuchungen über Zahl und Anordnung der Muskelspindeln in den Kaumuskeln des Menschen. Anat Anz 100:258–264
3. Godaux E, Desmedt JE (1975) Human masseter muscle: H and tendon reflexes. Arch Neurol 32:229–234
4. Hopf HC (1987) Masseter reflex in brain-stem tumors. Electroencephalogr Clin Neurophysiol 66:S47
5. Hopf HC, Gutman L (1990) Diabetic 3rd nerve palsy: evidence for a mesencephalic lesion. Neurology 40:1041–1045
6. Hopf HC, Thömke F, Gutmann L (1991) Midbrain versus pontine medial longitudinal fasciculus lesions: the utilization of masseter and blink reflexes. Muscle Nerve 14: 326–330
7. Kimura J, Rodnitzky RL, Van Allen MW (1970) Electrodiagnostic study of trigeminal nerve. Neurology 20:574–583
8. Lewis GR, Pilcher R, Yemm R (1980) The effect of stimulus strength on the jaw-jerk response in man. J Neurol Neurosurg Psychiatry 43:699–704
9. McIntyre AK, Robinson RG (1959) Pathway for the jaw jerk in man. Brain 82: 468–474
10. Ongerboer de Visser BW (1982) Afferent limb of the human jaw reflex: electrophysiologic and anatomic study. Neurology 32:563–566
11. Ongerboer de Visser BW, Goor C (1976) Jaw reflexes and masseter electromyograms in mesencephalic and pontine lesions. J Neurol Neurosurg Psychiatry 39:90–92
12. Sanders EACM, Ongerboer de Visser BW, Barendswaard EC, Arts RJHM (1985) Jaw, blink and corneal reflex latencies in multiple sclerosis. J Neurol Neurosurg Psychiatry 48:1284–1289
13. Struppler A, Struppler E, Adams R (1963) Local tetanus in man. Arch Neurol 8:162–178
14. Szentagothai J (1948) Anatomical considerations of monosynaptic reflex arcs. J Neurophysiol 11:445–454
15. Yates SK, Brown WF (1981) The human jaw jerk. Electrophysiologic methods to measure latency, normal values and changes in multiple sclerosis. Neurology 31:632–634

The Masseter Inhibitory Reflex in Pontine Lesions

B.W. Ongerboer de Visser and G. Cruccu

Departments of Clinical Neurophysiology and Neurology, University of Amsterdam, Academicsh Mediisch Centrum, Meibergdreef 9, 1105 AZ Amsterdam-Zuidoost, The Netherlands

A variety of stimuli anywhere within the mouth or on the facial skin of the maxillary or mandibular trigeminal division evoke bilateral inhibitory reflex responses in jaw-closing muscles. One of these reflexes is the masseter inhibitory reflex (MIR) [1–3], which consists of early (SP1)and late (SP2) phases of an electrical silent period in masseter voluntary activity (Fig. 1A). Characteristics of the two reflex components are standardized by giving a reflex threshold electrical stimulus three times to the mental nerve at the chin during maximal clenching of the jaws (Fig. 2). It may be necessary to use a concentric needle electrode instead of surface electrodes, particularly if the signal is contaminated by facial muscle activity. Maximal contraction has to be maintained for not more than 3 s with the aid of audiovisual feedback by the EMG apparatus. After relaxation, contraction is repeated every 30 s to avoid disappearance of EMG activity between the two silences.

The SP1 response, with a latency time of 10–15 ms, is probably mediated by medium-size myelinated afferents (A-beta) [4]. The SP2 response with a latency of 40–50 ms is probably mediated by afferents in the A-beta fiber range as well, although some authors regard it as a nociceptive reflex.

After stimulation of the mental nerve, impulses reach the pons via the sensory mandibular root of the trigeminal nerve (Fig. 1B) [2, 3]. For SP1, impulses enter the ipsilateral trigeminal tract and terminate at the level of the mid pons. They are then relayed by one inhibitory interneuron located close to the ipsilateral trigeminal motor nucleus. The inhibitory interneuron projects onto jaw-closing motoneurons bilaterally. The whole circuit lies in the mid pons. For SP2, afferents descend in the spinal trigeminal tract and connect with a polysynaptic chain of excitatory interneurons, probably located in the lateral reticular formation, at the level of the pontomedullary junction. The last interneuron of the chain is inhibitory and gives rise to ipsi- and contralateral collaterals that ascend medial to the right and left spinal trigeminal complexes, to reach the trigeminal motoneurons.

Inhibitory reflexes play a role in the reflex control of mastication, by preventing intraoral damage that could occur with uncontrolled contraction of jaw closing muscles, and in jaw movements during speech. These reflexes cannot be tested by purely clinical procedures. From the clinical point of

L.R. Caplan and H.C. Hopf (Eds.)
Brain-Stem Localization and Function

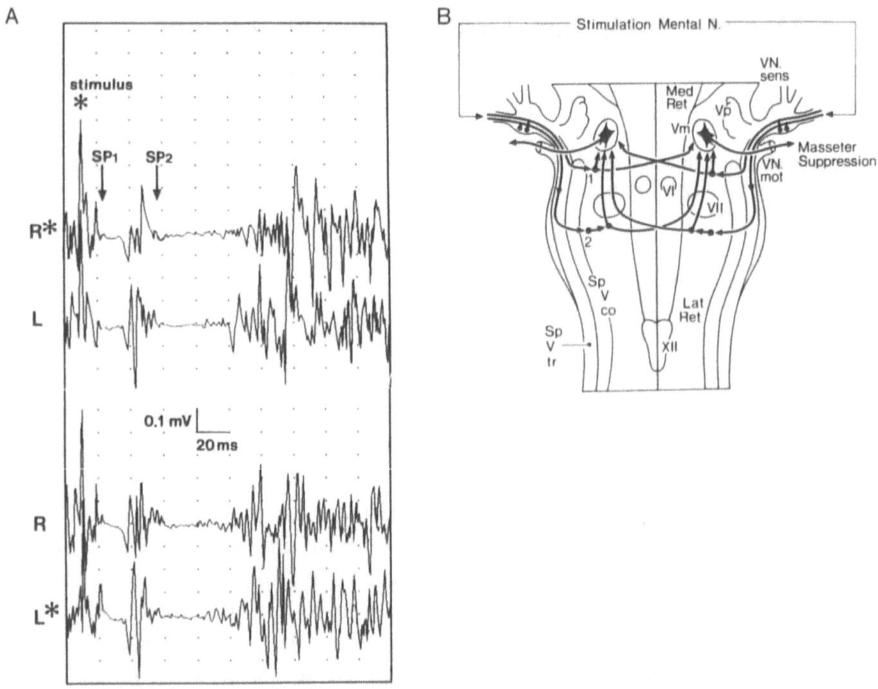

Fig. 1. A Example of normal masseter inhibitory reflex (MIR) with bilateral early (*SP1*) and late (*SP2*) silent periods. The upper two traces show responses from the right (*R*) and left (*L*) masseter muscle evoked by stimulation of the right (*) mental nerve. The lower two traces are the MIR responses obtained by stimulation of the left (*) mental nerve. **B** Presumed location of interneurons subserving the two components of the masseter ingibitory reflex: *1*, mid-pontine interneurons subserving the bilateral early component of electromyographical (EMG) silence (SP1); *2*, interneurons, probably located in the pontomedullary junction, subserving the bilateral late component of EMG silence (SP2). A lesion involving the Sp V tr with or without involvement of Sp V co would only interfere with the bilateral SP1 or SP2, both elicited by stimulation of the mental nerve on the affected side. A larger lesion extending more medially or a paramedian lesion sparing the Sp V co would also interfere with the SP1 or SP2 or both, ipsilateral to the affected side elicited by stimulation of the nerve on the intact side. A median lesion (depending on the extent of the lesion) would interfere with the SP1 or the SP2, both contralateral to the stimulation of the nerve on either side. In case of a median lesion, and depending on the extent of pathology, the lesion would interfere with the SP1 or the SP2 contralateral to the stimulation of the nerve on either side. Similar to the late blink [10] and corneal [11] reflexes, the ascending routes are thought to be located in the Lat Ret formation. *Vm*, trigeminal motor nucleus; *Vp*, trigeminal principal sensory nucleus; *Sp V tr*, spinal trigeminal tract; *Sp V co*, spinal trigeminal complex; *VI*, abducens nucleus; *VII*, facial nucleus; *XII*, hypoglossal nucleus; *VN mot*, trigeminal motor nerve; *VN sens*, trigeminal sensory nerve; *Med Ret*, medial reticular formation; *Lat Ret*, lateral reticular formation

view, the recording of the MIR can be of particular value because the results, in combination with findings of other trigeminal reflexes, may provide information on the functional implication of a CT or MRI lesion involving the pons. Furthermore, analysis of the SP1 and SP2 components

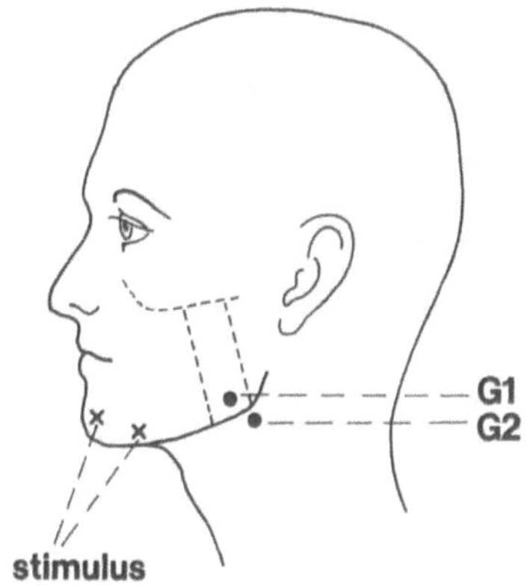

Fig. 2. Electrode placements for recording the masseter inhibitory reflex, with electrode placements for stimulating the mental nerve

helps in localizing the site of damage along the trigeminal paths. Indeed, in some patients it may be that no signs of trigeminal impairment are visible clinically, and other trigeminal reflex responses are even normal, yet MIR testing reveals trigeminal involvement.

Our studies on reflex pathways through the brain stem were performed in 16 patients with localized lesions [3]. Clinical symptoms, the types of pathology, and the location of the lesion in the brain stem as well as the results of the MIR examinations found in these patients are summarized in Table 1. Schematic representations of the MIR abnormalities are shown in Fig. 3. Patients 1 and 2 and patients 14, 15, and 16 had normal MIR findings. Their lesions were at the medullary or midbrain level, respectively. Conversely, the MIR showed abnormal features in all other patients with a pontine lesion or pathology confined to the pontomedullary junction. The pattern of abnormality of the MIR reflects the site of the brain-stem lesion. The lesion involves the dorsal pontine tegmentum at the level of the mid pons if both SP1 and SP2 are affected, and the lower pons or the pontomedullary junction if SP2 alone is affected.

In Fig. 3, type A represents an afferent type of abnormality; i.e., absent or incomplete or delayed direct and crossed responses to one-sided stimulation indicate a lesion along the afferent paths before the site of crossing of impulses. Noteworthy is the fact that SP1 is more susceptible than SP2 to extra-axial lesions, such as those in trigeminal neuropathy. A mixed type of abnormality (type B; Fig. 3), i.e., afferent incomplete EMG suppression, a delay or block to one sided-stimulation, plus abnormalities of the crossed responses alone or abnormal crossed responses to stimulation of either side (type D; Fig. 3), indicate a lesion within the brain stem. Figure 4 shows an example of the mixed type found in patient 6 (Fig. 3), with a brain-stem lesion in the left dorsal half at the pontomedullary level. In Fig. 5 an MRI lesion on the midline of the lower pons in patient 11 is seen (Fig. 3), who had abnormal late crossed responses. Type C (Fig. 3) represents the abnormality mostly seen in large pontine lesions where afferent and crossed routes

Table 1. Clinical findings, locations of brain-stem lesions demonstrated by magnetic resonance imaging (MRI), and abnormal findings from the recorded masseter inhibitory reflex (MIR) in the 16 patients with brain-stem lesions

Patient no., sex, age (years)	Symptoms	Brain-stem lesion		MIR	
		Pathology	Location (MRI)	SP1	SP2
1, M, 26	L trigeminal hypalgesia, L arm paresis	Tumor	Med Ob1–C4, L paramedian		
2, M, 19	L trigeminal neuralgia, nystagmus, dysphonia, R lateropulsion	Angioma	Med Ob1, L paramedian		
3, F, 67	R trigeminal hypalgesia	Infarction	Lower pons, R dorsolateral		Abn
4, M, 70	R trigeminal hypalgesia	Infarction	Pontomedullary junction, L dorsolateral		Abn
5, M, 40	L abducens and facial pareses	Tumor	Lower pons, R dorsolateral		Abn
6, F, 28	L unilateral internuclear ophthalmoplegia	Infarction	Pontomedullary junction, L dorsomedial		Abn
7, M, 49	R trigeminal neuralgia/ hypalgesia	Demyelination	Mid pons, R dorsolateral	Abn	Abn
8, F, 49	R trigeminal analgesia, R hearing loss, R hemiataxia	Infarction	Lower + mid pons + Med Ob1, R dorsolateral	Abn	Abn
9, F, 41	L trigeminal hypalgesia, L abducens and facial pareses, R hemiparesis	Tumor	Lower + mid pons, L half	Abn	Abn
10, F, 47	L > R abducens paresis, mild L facial paresis, nausea, vertigo, L hearing loss	Hemorrhage	Lower + mid pons, Midline	Abn	Abn
11, M, 45	Opsoclonus, mild R facial paresis	Arteriovenous malformation (AVM)	Lower pons, Midline		Abn
12, M, 53	L abducens paresis, R hemiparesis/ hypalgesia, vertigo	Hemorrhage	Upper + mid pons, L paramedian	Abn	Abn
13, F, 63	L trigeminal hypalgesia, L jaw muscles paralyses	Tumor	Upper + mid pons, L dorsolateral	Abn	Abn
14, M, 47	L arm dysmetria	AVM	Midbrain tegmentum, L half		
15, F, 52	R hemihypalgesia, R hemiataxia/hypotonia	Cyst (listerellosis)	Midbrain tegmentum, L paramedian		
16, M, 24	L ocular palsy/ptosis/ dilated pupil, R hemihypalgesia	Hemorrhage	Midbrain tegmentum, L paramedian		

R, on right side; L, on left side; MRI, magnetic resonance imaging; Med Ob1, medulla oblongata; MIR, masseter inhibition reflex; SP1, first silence; SP2, second silence; Abn, abnormal; normal is not indicated.

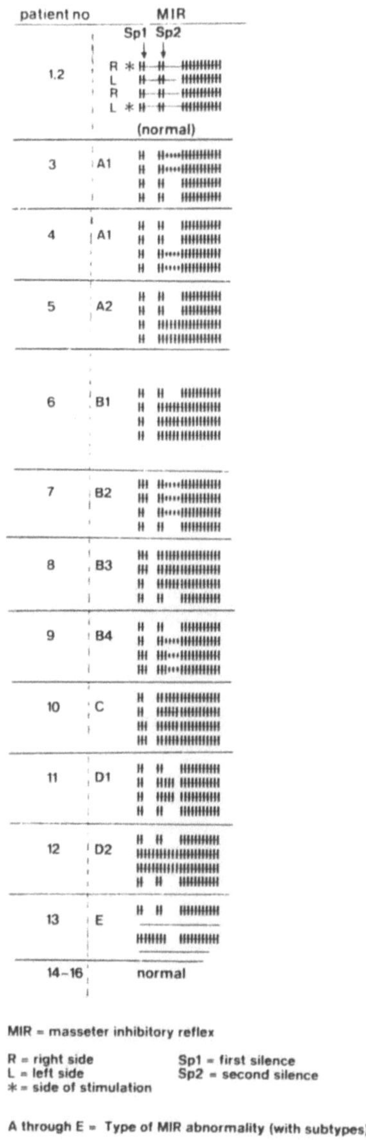

Fig. 3. Reflex patterens of the masseter inhibitory reflex (*MIR*) that we observed in 16 patients with a localized lesion in the brain stem (see also Table 1). The first two patients had medullary lesions with normal SP1 and SP2 configurations and latencies. Patients 14–16 had unilateral midbrain lesions. They also had normal MIR responses. Patients 3–13 had lesions located in the pons or the pontomedullary junction. In these patients patterns of five types (*A–E*) of abnormal MIR components were demonstrated. The subtypes were characterized by a complete or incomplete suppression or a delay of inhibition of masseter EMG activity

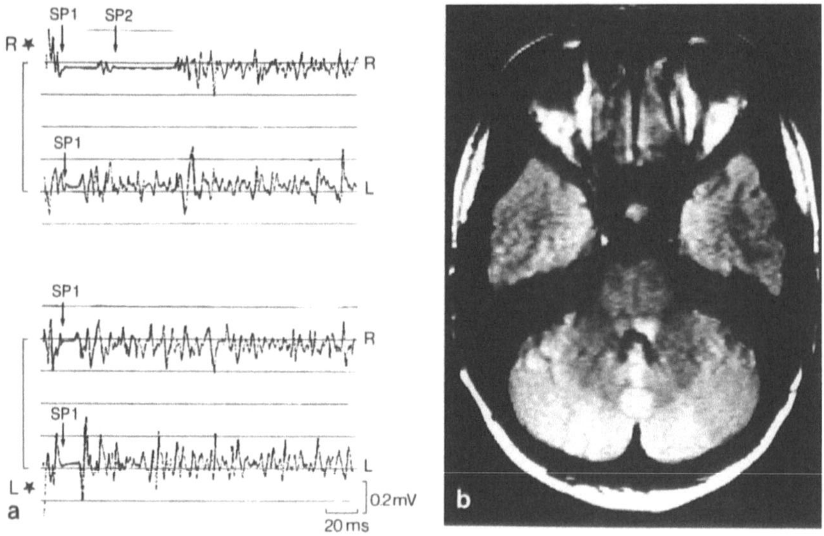

Fig. 4. a A mixed-type abnormality of the masseter inhibitory reflex in patient 6 (Table 1 and Fig. 3). *Upper* pair of *traces*: normal SP1 and SP2 on the right (*R*) and a normal SP1 with an absent SP2 on the left (*L*) side after stimulation of the right mental nerve (*R**). *Lower* pair of *traces*: normal bilateral SP1 with an absent bilateral SP2 after stimulation of the left mental nerve (*L**). **b** Magnetic resonance imaging (MRI) scan (T_1 weighted) of patient 6, showing a small area of increased signal, probably an infarction in the left dorsal half at the level of the pontomedullary junction

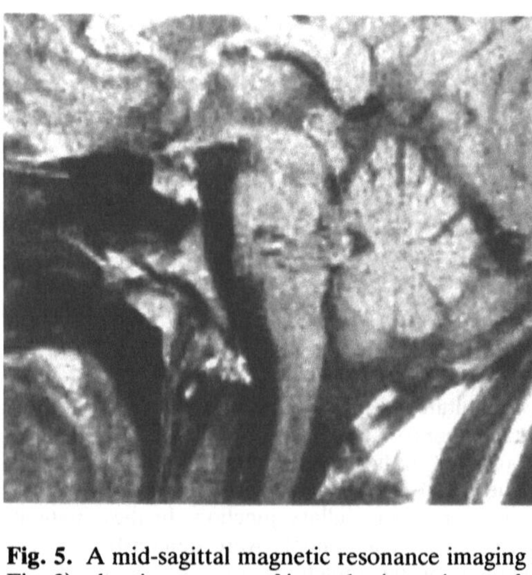

Fig. 5. A mid-sagittal magnetic resonance imaging (MRI) scan of patient 11 (Table 1 and Fig. 3), showing an area of irregular intensity, probably an arteriovenous malformation on the midline of the lower pons, just rostral to the pontomedullary junction and extending into the fourth ventricle. After stimulation of the mental nerve, pattern D of reflex abnormality (Fig. 3) was seen

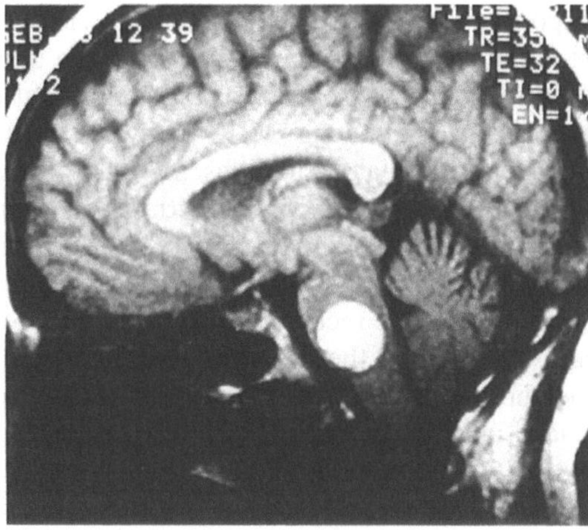

Fig. 6. A mid-sagittal magnetic resonance imaging (MRI) scan of patient 10 (Table 1 and Fig. 3), showing a large hemorrhage confined to the lower and mid pons. After stimulation of the mental nerve, pattern C of reflex abnormality (Fig. 3) was seen

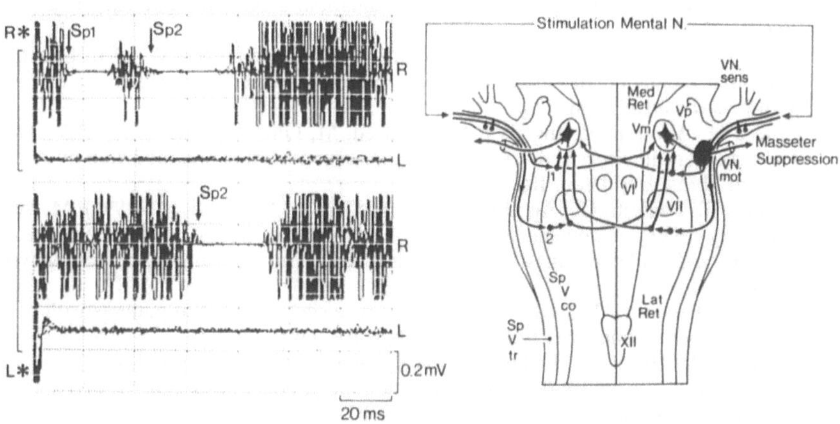

Fig. 7. *Left*: Pattern E of reflex abnormality (Fig. 3) was found in patient 13 (Table 1). Stimulation on the normal, right (*R*) side (*R**) yielded normal SP1 and SP2, while a total loss of masseter EMG activity occurred on the left (*L*) side, the paralytic side. A stimulus on the affected side (*L**) evokes only the contralateral, second inhibitory reflex (SP2) with delay due to compression on the afferent path. *Right*: Location of the lesion. Pathology was situated in the dorsolateral portion of the mid pons and caused left-sided trigeminal hypalgesia and paralysis of jaw muscles

are affected. An example of such a lesion is presented in Fig. 6. Figure 7 shows type-E abnormality in a patient with a lesion located in the dorsolateral portion of the upper and mid pons. There is no EMG activity on the side of masseter paralysis, and SP1 is absent only upon stimulation on that side. An efferent type of abnormality, i.e., absent, delayed, or incomplete responses confined to the masseter muscle on one side, regardless of the side of stimulation, is extremely rare except in unilateral masticatory spasms [5].

In this paper we have discussed abnormal MIR features caused by structural lesions involving MIR pathways through the pontine tegmentum. However, MIR abnormalities may also occur in functional disorders and dysfunction of motor control without anatomical lesions of the brain stem. Examples are the abnormalities that can be seen in temporomandibular dysfunction [6, 7] and diffuse sclerosing osteomyelitis of the mandible [8]. Furthermore, the recovery cycle of the SP2 component studied by delivering paired stimuli to the mental nerve has been found to be greatly enhanced and most sensitive in Parkinson's disease and dystonia involving the craniofacial muscles [5, 9].

References

1. Godeaux E, Desmedt JE (1975) Exteroceptive suppression and motor control of the masseter and temporalis muscles in normal man. Brain Res 85:447–458
2. Ongerboer de Visser BW, Goor C (1976) Cutaneous silent period in masseter muscles: a clinical and electrodiagnostic evaluation. J Neurol Neurosurg Psychiatry 39:674–679
3. Ongerboer de Visser BW, Cruccu G, Manfredi M, Koelman JHTM (1990) Effects of brainstem lesions on the masseter inhibitory reflex: functional mechanisms of reflex pathways. Brain 113:781–792
4. Cruccu G, Agostino R, Inghilleri M, Manfredi M, Ongerboer de Visser BW (1989) The masseter inhibitory reflex is evoked by innocuous stimuli and mediated by A beta afferent fibres. Exp Brain Res 77:447–450
5. Cruccu G, Pauletti G, Agostino R, Berardelli A, Manfredi M (1991) Masseter inhibitory reflex in movement disorders, Huntington's chorea, Parkinson's disease, dystonia and unilateral masticatory spasms. Electroencephalogr Clin Neurophysiol 81:24–30
6. Skiba TJ, Laskin DM (1981) Masticatory muscle silent periods in patients with MPD syndrome before and after treatment. J Dent Res 60:699–706
7. Sharav Y, McGrath PA, Dubner R (1982) Masseter inhibitory periods and sensations evoked by electrical tooth pulp stimulation in patients with oral-facial pain and mandibular dysfunction. Arch Oral Biol 27:305–310
8. Groot RH, Ongerboer de Visser BW, van Merkesteyn JPR, Speelman JD, Bras J (1992) Changes in masseter inhibitory reflex responses in patients with diffuse sclerosing osteomyelitis of the mandible. Oral Surg Oral Med Oral Pathol 74:727–732
9. Ongerboer de Visser BW, Cruccu G (1993) Neurophysiological examination of the trigeminal, facial, hypoglossal, and spinal accessory nerves in cranial neuropathies and brainstem disorders. In: Brown WF, Bolton CF (eds) Clinical electromyography, 2nd edn. Butterworth Boston
10. Ongerboer de Visser BW, Kuypers HGJM (1978) Late blink reflex changes in lateral medullary lesions. An electrophysiological and neuroanatomical study of Wallenberg's syndrome. Brain 101:285–294
11. Ongerboer de Visser BW, Moffie D (1979) Effects of brain-stem and thalamic lesions on the corneal reflex. An electrophysiological and anatomical study. Brain 102: 595–608

Diagnostic Localizing Value of the Electrically Elicited Blink Reflex

G. Krämer

Neurologische Klinik, Universität Mainz, Langenbeckstraße 1, 55131 Mainz, Germany

Introduction

The blink reflex (BlinkR) following electrical stimulation of the supraorbital nerve consists of an early ipsilateral reflex component (R1) and bilateral late reflex components ipsilaterally (R2) and contralaterally (R2c). An additional, and even later component, also bilateral, R3, R3c was first described by Penders and Delwaide [34] and so far has not been studied systematically.

Anatomy and Physiology

The common afferent limbs of the BlinkR are sensory fibers of the trigeminal nerve, whereas the common efferent limb is the facial nerve. The central loop of the R1 reflex arc runs from the lateral mid-pons to the ipsilateral facial nucleus, in particular the intermediate subnucleus. There is clinical evidence that it runs through, or close to, the trigeminal principal sensory nucleus [16, 29, 42]. R1 does not habituate [30], since it involves only two or three synapses. With facilitation by conditioning stimulation of other cutaneous afferents, a contralateral R1 can be observed, indicating an additional crossed oligosynaptic trigeminofacial connection [45, 48].

The central loops of the R2 reflex components are relayed through a more complex polysynaptic route involving the pons and lateral medulla [33]. After entering the pons, the fibers descend with the trigeminal spinal tract and nucleus to the caudal medulla. Connections to the facial nuclei ascend ipsilaterally and, after crossing at the caudal level, course contralaterally within the lateral propriobulbar system of the reticular formation [12, 13]. Clinical correlations favor a more lateral position of the ipsilateral as compared to the contralateral ascending fibers [43]. Habituation of the R2 response takes place within the interneuron pool around the facial nucleus [36]; the R1 and R2 responses are cutaneous and nociceptive in origin, and R2 correlates with blinking of the lids, whereas no lid movement is observed with R1 [8].

L.R. Caplan and H.C. Hopf (Eds.)
Brain-Stem Localization and Function

Although little is known about the anatomical and functional organization of the R3 response, its central circuit is undoubtedly polysynaptic. It shows a more pronounced habituation than R2, seems to also be nociceptive, and is possibly pain related [4, 40].

Technical Aspects and Evaluation

The response of the orbicularis oculi muscle is recorded bilaterally by surface electrodes placed over the lower lid halfway between the inner and outer edge of the orbit. The reference electrode is placed over the lateral aspect of the nose or over the temple, and the ground electrode, on the forehead, under the chin, or around the arm. The supraorbital branch of the trigeminal nerve is stimulated with the cathode placed over the supraorbital foramen. Single electrical shocks of 0.1–0.2 ms duration are applied with increasing intensity (3–20 mA) until maximal and stable R1 and R2 responses are obtained. At least five successive reflexes are then evaluated and the mean latencies are calculated for each component. Habituation is avoided by stimulating at random intervals of 10–20 s.

R3 is not detected in routine studies, because it reproducibly occurs only after stronger stimuli which are just below the pain threshold, i,e., five to six times the individual stimulus threshold [40]. Sometimes R3 merges with the end of R2 or begins after a pause of electric silence 70–90 ms after the stimulus. Voluntary activation and conditioning stimulation using subliminal intensities and intervals of about 50 ms as well as paired stimuli of above-threshold intensities at 5 ms intervals elicits or facilitates R1 responses otherwise not reproducible [24, 38].

Normal values. The BlinkR latencies change little during a person's lifetime. In full-term newborns, R1 latencies are between 12.1 ± 1.1 ms [19] and 12.8 ± 0.6 ms [10]. R2 and R2c components are not observed in about 50% of premature infants. By the time of normal delivery, the R2 latency is about 10 ms longer than in adults, and in the age group between 7 and

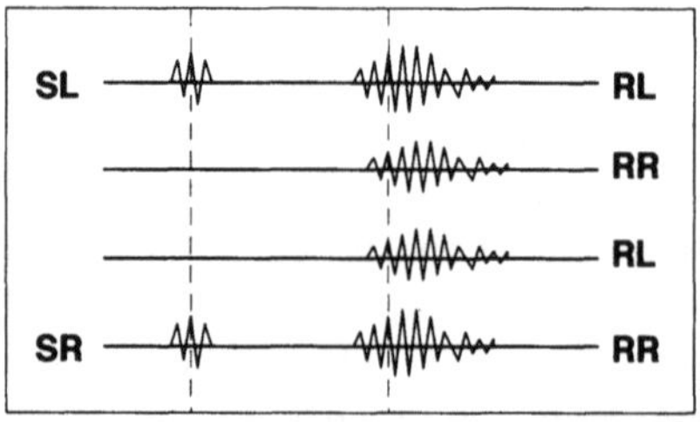

Fig. 1. Normal pattern of the electrically elicited blink reflex. *SL*, stimulation on the left side; *SR*, stimulation on the right side; *RL*, recording on the left side; *RR*, recording on the right side; *vertical dotted lines*, upper limits of the latencies

12 years it is 5 ms longer [10]. In adults, the upper latency limits (mean + 2.5 SEM) are about 12 ms for R1 and about 40 ms for R2 (see also Fig. 1). Our own data are 12.1 ms for R1, 39.5 ms for R2, and 41.5 ms for R2c. There are no reliable differences between the age groups of 20–39 years, 40–59 years, and 60–79 years (unpublished data).

The criteria for considering R1 as abnormal are unilateral loss or delay of at least 1.4 ms and unilateral or bilateral delay above the upper limit (mean + 2.5 SEM). The criteria for the R2 components are differences of 5 ms or more between R2 and R2c following unilateral stimulation and side differences of 7 ms or more between the left and right R2 and/or R2c following bilateral stimulation [24].

Clinical Implications

The latencies of the individual BlinkR components reflect conduction along the entire length of the supraorbital and facial nerves and their intraaxial central connections. Abnormal latencies may be due to lesions at any site of the reflex arc and indicate brain-stem or peripheral nerve disease.

Peripheral Lesions

Afferents. In trigeminal sensory neuropathy as well as surgically treated trigeminal neuralgia, an afferent type R1 delay is observed on the affected side [3, 6, 11, 27, 47]. A corresponding R2 delay of up to 5 ms may, however, escape detection because of the greater variability of R2 latencies (Fig. 2).

Efferents. In severe Bell's palsy and related diseases, there is an efferent type of abnormality characterized by a delay or loss of R1, R2, and R2c on the affected side following stimulation of either side (Fig. 3). These abnormalities are not necessarily apparent at the onset of the palsy, but invariably develop within the first week. Prognosis is favorable if the reflex

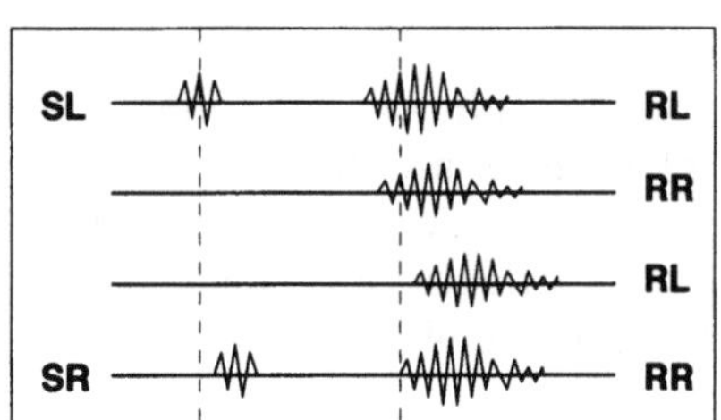

Fig. 2. Right afferent lesion with delay of R1, R2, and R2c (for abbreviations see Fig. 1)

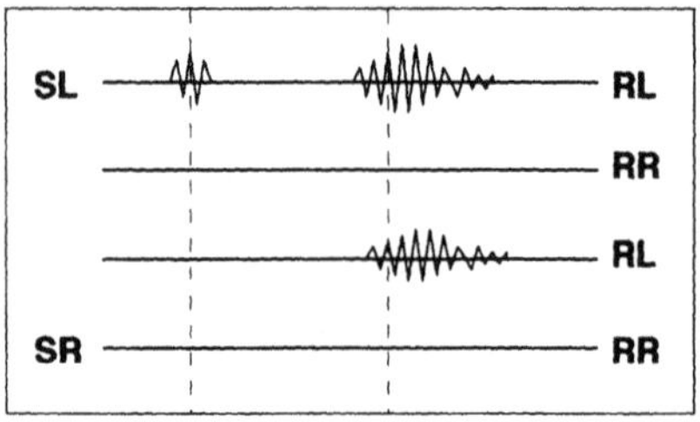

Fig. 3. Right efferent lesion with loss of R1 and R2 after ipsilateral stimulation and loss of R2c after contralateral stimulation (for abbreviations see Fig. 1)

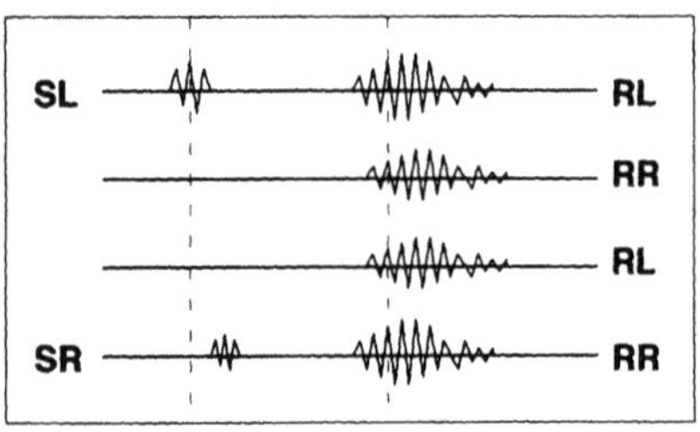

Fig. 4. Right pontine lesion with isolated delay of R1 (for abbreviations see Fig. 1)

components show only a delay or reappear within 10 days. Prognosis is poor if the responses remain absent for more than 3 weeks. In such patients, a R1 response can be recorded from perioral, mental, or frontal muscles following aberrant regeneration [25]. In hemifacial spasm, there may be increased R1 amplitudes and synkinetic responses of R1 and R2 in the mental muscles [24]. Patients with Guillain-Barré syndrome, Fisher syndrome, or hereditary motor and sensory neuropathy type I (HMSN) may show a marked bilateral delay of R1, whereas corresponding R2 delays are less obvious due to the larger normal range [23].

Central Lesions

Pons. BlinkR abnormalities are common in patients with pontine lesions. Isolated absence or delay of R1 with normal R2 are reliable signs of focal pontine pathology indicating a lateral mid-pontine lesion [15, 29]. Isolated R1 changes are mainly found in patients with multiple sclerosis [20, 29], which may be associated with internuclear ophthalmoplegia [16, 24]. In our experience, less common possible causes include ischemia, tumor, and trauma (Fig. 4).

Medulla. Various patterns of R2 abnormalities have been observed in medullary lesions (Figs. 5, 6). However, their diagnostic localizing value is

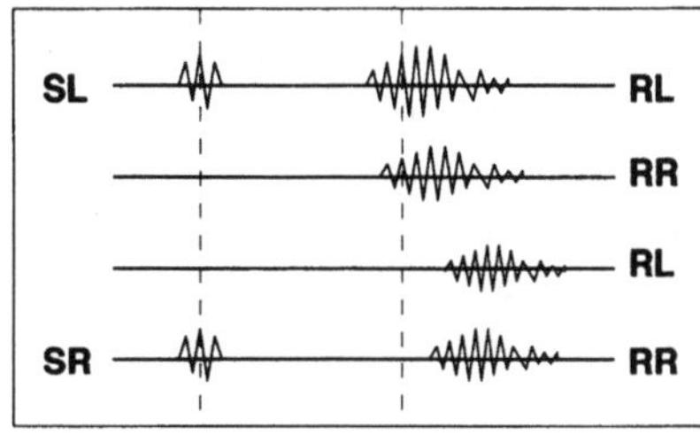

Fig. 5. Right lateral medullary lesion with normal R1 and delay of R2 and R2c (for abbreviations see Fig. 1)

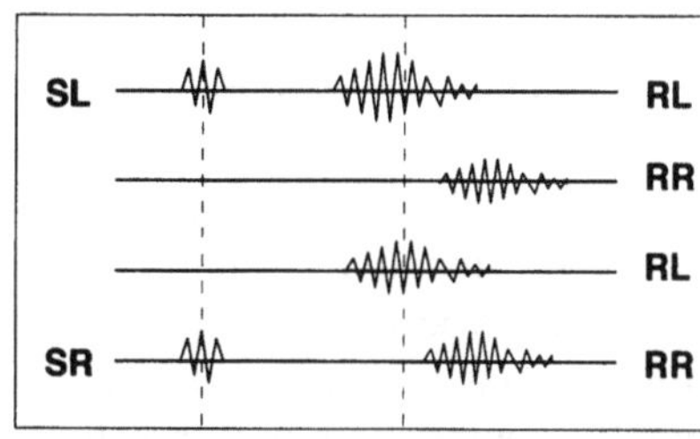

Fig. 6. Right paramedian medullary lesion with normal R1, ipsilateral delay of R2, and contralateral delay of R2c (for abbreviations see Fig. 1)

limited due to larger variation of the normal range and strong supraseg-mental influences, particularly from the lower postcentral region. This causes delay or suppression of the R2 and R2c responses which is not counteracted by paired stimulation [1, 26, 32]. Isolated changes of R2c with normal R1 and R2 are attributed to very small lesions in the lateral medulla [43]. In Wallenberg's syndrome, infarction is usually accompanied by ipsilateral delay or loss of R2 and R2c components or ipsilateral loss of R2 and bilateral loss of R2c. In patients with pure medullary lesions, R1 is spared [33]. In comatose patients resuscitated from cardiac arrest, the R1 component reoccurs earlier than R2. Both R1 and R2 may be present in patients with apallic syndrome (vegetative state), but are abolished in brain death [18].

Pontomedullary Junction. Patients with multiple sclerosis very commonly have combined R1 and R2 abnormalities. In such patients, BlinkR abnormalities can be due to either larger or multiple lesions [22, 28].

Upper Cervical Level. Three out of four patients with cervical myelopathy due to mid-cervical compression at C 4/5 showed unilateral or bilateral abnormalities of the BlinkR R3 component. The R1 and R2 responses always remained normal [39]. These preliminary findings await verification.

Suprasegmental Lesions. A suprasegmental effect on R1 resulting in reduced excitability [17] and increased latency [7] was observed only in the acute stages of hemispheric lesions. This is recognized if it can be counteracted by facilitating maneuvers [35] and paired stimulation [25]. The R2 responses are under obvious activating suprasegmental influence. Therefore, hemispheric and mesencephalic lesions involving the lower postcentral connections may cause delay or loss of the R2 components on the clinically affected side [1, 5, 26, 32] and bilaterally [2].

Systemic Disease

BlinkR changes also occur in systemic disease, but are of no localizing value. Slight, but definitely delayed (as compared to age-matched healthy subjects), R1 and R2 responses were found in Huntington's disease and Gilles de la Tourette's syndrome [41]. An increased R1 amplitude associated with a shortened R2 latency and depressed R2c response has been interpreted as hyperexcitability of uncrossed interneurons and orbicularis oculi motoneurons in the Lesch-Nyhan syndrome [9]. In blepharospasm, the R1 and R2 components may not be separated and a contralateral R1 may be observed [37]. A BlinkR showing four components has been observed in periodic ataxia. Latencies of the unilateral first component corresponded to the normal R1 and those of the bilateral third and fourth components were both in the range of the normal R2. An additional second component had a latency of about 21 ms which did not habituate to 10/s stimulation [49]. Sedatives, particularly benzodiazepines and barbiturates, totally or partially abolish the late blink reflex components [4, 21, 31]. Narcotics such as fentanyl have little effect on R1 and R2, but almost abolish R3, which reappears after naloxone administration [4].

Conclusions

The electrically elicited blink reflex is a noninvasive and easily applicable brain-stem reflex with wide clinical applicability. Of the two components, R1 is more reliable, and isolated R1 changes have the highest diagnostic localizing value. Magnetic resonance imaging may fail to demonstrate some small or reversible lesions [46]. In such cases, combined electrophsiological multireflex testing is recommended [14, 44]. Potential suprasegmental influences as well as effects of systemic disease and disturbances have to be taken into account.

References

1. Berardelli A, Accornero N, Cruuci G, Fabiano F, Guerrisi V, Manfredi M (1983) The orbicularis oculi response after hemispheral damage. J Neurol Neurosurg Psychiat 46:837–843
2. Csécsei G (1980) Alteration of the blink reflexes in hemispheral lesions. Electromyogr Clin Neurophysiol 20:141–152
3. Cruccu G, Agostino R, Fornarelli M, Inghilleri M, Manfredi M (1987) Neurophysiological assessment of trigeminal function after surgery for trigeminal neuralgia. Neurology 37:631–638
4. Cruccu G, Ferracuti S, Leardi MG, Fabbri A, Manfredi M (1991) Nociceptive quality of the orbicularis oculi reflexes as evaluated by distinct opiate- and bezodiazepine-induced changes in man. Brain Research 556:209–217
5. Dengler R, Kossev A, Gippner C, Struppler A (1982) Quantitative analysis of blink reflexes in patients with hemiplegic disorders. Electroenceph Clin Neurophysiol 53:513–524
6. Feldman RG, Niles C, Proctor SP, Jabre J (1992) Blink reflex measurement of effects of trichloroethylene exposure on the trigeminal nerve. Muscle & Nerve 15:490–495
7. Fisher MA, Shahani BT, Young RR (1979) Assessing segmental excitability after acute rostral lesions: the blink reflex. Neurology 29:45–50
8. Gregoric M (1973) Habituation of the blink reflex. In: Desmedt JE (ed) New developments in electromyography and clinical neurophysiology, vol 3. Karger, Basel, pp 673–677
9. Hatanake T, Higashino H, Woo M, Yasuhara A, Sugimoto T, Kobayashi Y (1990) Lesch-Nyhan syndrome with delayed onset of self mutilation: hyperactivity of interneurons at the brainstem and blink reflex. Acta Neurol Scand 81:184–187
10. Hatanaka T, Yasuhara A, Kobayashi Y (1990) Electrically and mechanically elicited blink reflexes in infants and children – maturation and recovery curves of blink reflex. Electroenceph Clin Neurophysiol 76:39–46
11. Hess K, Kern S, Schiller HH (1984) Blink reflex in trigeminal sensory neuropathy. Electromyogr Clin Neurophysiol 24:185–190
12. Holstege G, Kuypers HGJM (1977) Propriobulbar fibre connections to the trigeminal, facial and hypoglossal motor nuclei: I. An anterograde degeneration study in the cat. Brain 100:239–264
13. Holstege G, Kuypers HGJM, Dekker JJ (1977) The organization of the bulbar fibre connections to the trigeminal, facial and hypoglossal nuclei: II. An autoradiographic tracing study in the cat. Brain 100:265–286
14. Hopf HC (1993) Topodiagnostic value of brain stem reflexes. Muscle & Nerve 16 (in press)
15. Hopf HC, Tettenborn B, Krämer G (1990) Pontine supranuclear facial palsy. Stroke 21:1754–1757
16. Hopf HC, Thömke F, Gutmann L (1991) Midbrain vs. pontine medial longitudinal fasciculus lesions: the utilization of masseter and blink reflexes. Muscle & Nerve 14:326–330
17. Kaplan PE, Kaplan C: Blink reflex: review of methodology and its application to patients with stroke syndrome. Arch Phys Med Rehabil 61:30–33
18. Kano T, Shimoda O, Morioka T, Yagishita Y, Hshiguchi A (1992) Evaluation of the central nervous function in resuscitated comatose patients by multilevel evoked potentials. Resuscitation 23:235–248
19. Khater-Boidin J, Duron B (1987) The orbicularis oculi reflex in healthy premature and full-term newborns. Electroenceph Clin Neurophysiol 67:479–484
20. Kiers L, Carroll WM (1990) Blink reflexes and magnetic resonance imaging in focal unilateral central trigeminal pathway demyelination. J Neurol Neurosurg Psychiat 53:526–529
21. Kimura J (1973) The blink reflex as a test for brain-stem and higher central nervous function. In: Desmedt JE (ed) New developments in electromyography and clinical neurophysiology, vol 3. Karger, Basel, pp 682–691

22. Kimura J (1975) Electrically elicited blink reflex in diagnosis of multiple sclerosis. Brain 98:413–426
23. Kimura J (1983) Clinical uses of the electrically elicited blink reflex. In: Desmedt JE (ed) Motor control mechanisms in health and disease (Advances in neurology, vol 39). Raven Press, New York, pp 773–786
24. Kimura J (1989) The blink reflex. In: Kimura J (ed) Electrodiagnosis in diseases of nerve and muscle: principles and practice, 2nd edn. Davis, Piladelphia, pp 307–331
25. Kimura J, Rodnitzky RL, Okawara SH (1975) Electrophysiologic analysis of aberrant regeneration after facial nerve paralysis. Neurology 25:989–993
26. Kimura J, Wilkinson JT, Damasio H, Adams HR Jr, Shivapour E, Yamada T (1985) Blink reflex in patients with hemispheric cerebrovascular accident (CVA). J Neurol Sci 67:15–28
27. Kimura J, Yamada T, Nitz DM (1980) Blink reflex as a test for lesions of the trigeminal nerve. Muscle & Nerve 3:184–188
28. Krämer G, Lowitzsch K, Lüder G, Kempkes K (1983) Blink reflexes in MS patients. In: Pedersen E, Clausen J, Oades L (eds) Actual problems in MS research. Fadl's Forlag, Kopenhagen-Aarhus-Odense, pp 45–49
29. Krämer G, Lowitzsch K, Besser R, Hopf HC (1986) Isolierte R1-Veränderungen des Blinkreflexes. In: Lowitzsch K (ed) Hirnstammreflexe. Methodik und klinische Anwendung. Thieme, Stuttgart-New York, pp 124–131
30. Lowitzsch G, Lüder G (1985) Habituation of the blink reflex: computer assisted quantitative analysis. Electroenceph Clin Neurophysiol 60:525–531
31. Marelli RA, Hillel AD (1989) Effects of general anasthesia on the human blink reflex. Head & Neck 11:137–149
32. Ongerboer de Visser BW (1983) Comparative study of corneal and blink reflex latencies in patients with segmental or with cerebral lesions. In: Desmedt JE (ed) Motor control mechanisms in health and disease (Advances in neurology, vol 39). Raven Press, New York, pp 757–772
33. Ongerboer de Visser BW, Kuypers HGJM (1978) Late blink reflex changes in lateral medullary lesions. An electrophysiological and neuro-anatomical study of Wallenberg's syndrome. Brain 101:285–294
34. Penders CA, Delwaide PJ (1973) Physiologic approach to the human blink reflex. In: Desmedt JE (ed) New developments in electromyography and clinical neurophysiology, vol 3. Karger, Basel, pp 649–657
35. Raffaele R, Palmeri A, Ricca G, Casabona A, Perciavalle V (1990) Unilateral enhancement of early and late blink refelx components in hemidystonia. Electromyogr Clin Neurophysiol 30:469–473
36. Rimpel J, Gryer D, Hopf HC (1982) Changes in the blink responses to combined trigeminal, acoustic and visual repetetive stimulation, studied in the human subject. Electroenceph Clin Neurophysiol 54:552–560
37. Roggenkämper P, Laskawi R, Damenz W, Schröder M (1991) Veränderungen des Blink-Reflexes bei Patienten mit essentiellem Blepharospasmus und Spasmus hemifacialis. Fortschr Ophthalmol 88:411–415
38. Rossi A, Scarpini C (1992) Gating of trigemino-facial reflex from low-threshold trigeminal and extratrigeminal cutaneous fibres in humans. J Neurol Neurosurg Psychiat 55:774–780
39. Rossi B, Pasca SL, Sartucci F, Siciliano G, Murri L (1989) Trigemino-cervical reflex in pathology of the brain stem and of the first cervical cord segments. Electromyogr Clin Neurophysiol 29:67–71
40. Rossi B, Risaliti R, Rossi A (1989) The R3 component of the blink reflex in man: a reflex response induced by activation of high threshold cutaneous afferents. Electroenceph Clin Neurophysiol 73:334–340
41. Smith SJM, Lees AJ (1980) Abnormalities of the blink reflex in Gilles de la Tourette syndrome. J Neurosurg Psychiat 52:895–898
42. Straschill M (1980) Orbicularis oculi Reflex mit fehlender früher Komponente und normaler später Reaktion. Z EEG-EMG 11:19–20

43. Tackmann W, Ettlin T (1982) Blink reflexes eleicited by electrical, acoustic and visual stimuli. Their relation to visual-evoked potentials and auditory brain stem evoked potentials in the diagnosis of multiple sclerosis. Eur Neurol 21:263–268
44. Tettenborn B, Caplan LR, Krämer G, Hopf HC (1992) Electrophysiology in ischemic brain stem disease. In: Berguer R, Caplan LR (eds) Vertebrobasilar arterial disease. Quality Medical Publishing, St. Louis:124–129
45. Trontelj MA, Trontelj JV (1978) Reflex arc of the first component of the human blink reflex: a single motoneurone study. J Neurol Neurosurg Psychiat 41:538–547
46. Uncini A, Faricelli A, Assetta M, Serio A, Tartaro A, Gambi D (1990) Electrophysiological and magnetic resonance imaging correlates of brainstem demyelinating lesions. Electromyogr Clin Neurophysiol 30:233–238
47. Valls-Sole J, Graus F, Font J, Pou A, Tolosa ES (1990) Normal proprioceptive trigeminal afferents in patients with Sjögren's syndrome and sensory neuropathy. Ann Neurol 28:786–790
48. Willer JC, Boulu P, Bratzlavsky M (1984) Electrophysiological evidence for crossed oligosynaptic trigemino-facial connections in normal man. J Neurol Neurosurg Psychiat 47:87–90
49. Yokota T, Hayashi H, Hirose K, Tanabe H (1990) Unusual blink reflex with four components in a patient with periodic ataxia. J Neurol 237:313–315

Facial Nerve Function and Taste

Magnetically Evoked Corticofacial Potentials of Orbicularis Oculi Muscle Conditioned by the Electrical Blink Reflex

J. Faig, M. Tegenthoff, and J.-P. Malin

Neurologische Klinik und Poliklinik, BG-Krankenanstalten Bergmannsheil, Ruhr-Universität, Gilsingstraße 14, 44789 Bochum, Germany

Introduction

A recent anatomical study [5] concerning corticofacial pathways in the monkey has demontrated mainly indirect, polysynaptic connections to the subnucleus of the facial nerve (FN) innervating upper facial muscles. These indirect corticofacial projections are mediated via interneurons of the reticular formation adjacent to the FN. In contrast, FN motor neurons supplying lower facial muscles receive only direct corticofacial innervation [5]. These observations were obtained using the technique of retro- and anterograde transport of horseradish peroxidase.

The terminal distributions of corticofacial pathways in man, however, have been examined only once; that study suggested direct connections on the basis of nerve degeneration [8].

These anatomical results led us to consider an electrophysiological method for examining the nature of the corticofacial projections to the upper facial muscles, in particular the orbicularis oculi muscle (OOM). The motor neurons of this muscle represent the common efferent path of two established neurophysiological methods with different afferent inputs to the FN, namely the electrically elicited blink reflex (BR) [9] and the corticofacial potentials of the OOM evoked by magnetic stimulation of the brain (briefly: MEP) [1, 10].

The late BR responses are known to be mediated polysynaptically through the reticular formation [3]. Experiments of habituation of late BR responses indicate the existence of a pool of interneurons – adjacent to the FN – as the common final relay for different sensory afferents of the BR [11, 13].

The aforementioned studies provide the basis for our question: Is there a functional linkage between the late BR responses and the corticofacial pathway to the OOM at the level of interneurons adjacent to the FN?

L.R. Caplan and H.C. Hopf (Eds.)
Brain-Stem Localization and Function

Method

Basis

Paired stimuli to the supraorbital nerve result in inhibition of late BR responses of the test reflex at interstimulus intervals of about 40 up to at least 800 ms [6]. This inhibition is due to reduced excitability of the interneurons caused by the conditioning stimulus [6, 13]. In the same way, a conditioning BR will inhibit a test MEP if there really is the assumed functional linkage at the interneuronal level.

Paired Stimulation Technique

The examination was carried out in seven healthy volunteers, aged 20–34 years.

Conditioning stimulus: electrical square wave impulses, left supraorbital nerve, 20–30 mA, 0.2 ms, stimulator of Neuropack 4 MEM-4104 G device, NIHON KOHDEN.

Test stimulus: magnetic stimulus over each hemisphere; optimal coil position was with the center of the coil approximately 4 cm lateral to Cz,

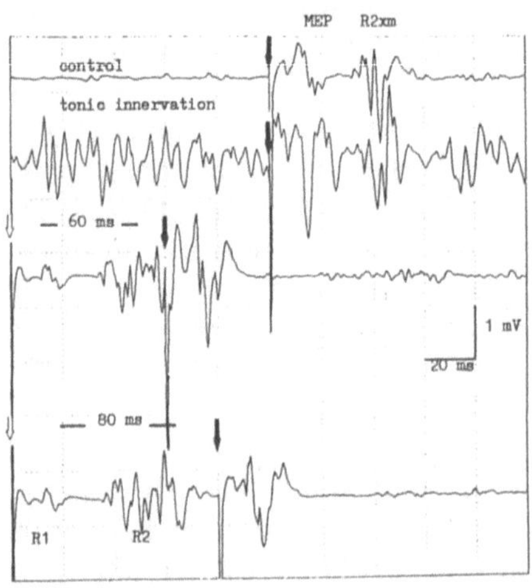

Fig. 1. Recording from left orbicularis oculi muscle (LOO); magnetic stimuli over the right hemisphere (*closed arrows*). (*Top to bottom*) *first trace*: single magnetic stimulation (*closed arrow*) LOO at rest, double discharge of corticofacial potential (*MEP*), and magnetically evoked late BR response (*R2xm*); *second trace*: facilitation by tonic innervation of target muscle allowing identification of MEP; *last two traces*: paired stimulation BR, elicited by the conditioning stimulus (*open arrow*); facilitation of MEP and simultaneous inhibition of R2xm following the second stimulus (*closed arrow*) at interstimulus intervals of 60 ms and 80 ms

appropriate direction of coil current [1, 10], circular coil, 50%–65% of maximal output (Magstim 200, NOVAMETRIX).

Interstimulus interval (ISI): between 2 and 300 ms, triggered by Programmer D 100, DIGITIMER.

Recording: (neuropack 4) lower eyelid against nasal reference on each side, monopolar needle electrodes in two, surface electrodes in five subjects; bandwith 50 Hz–3 kHz; appropriate sensitivity and time base.

Examination procedure: Step 1 – searching for the optimal position of the magnetic coil to evoke a reproducible control recording with the target muscle at rest; only contralateral potentials were evaluated. Step 2 – for the purpose of reliable identification of the corticofacial potential, recording at

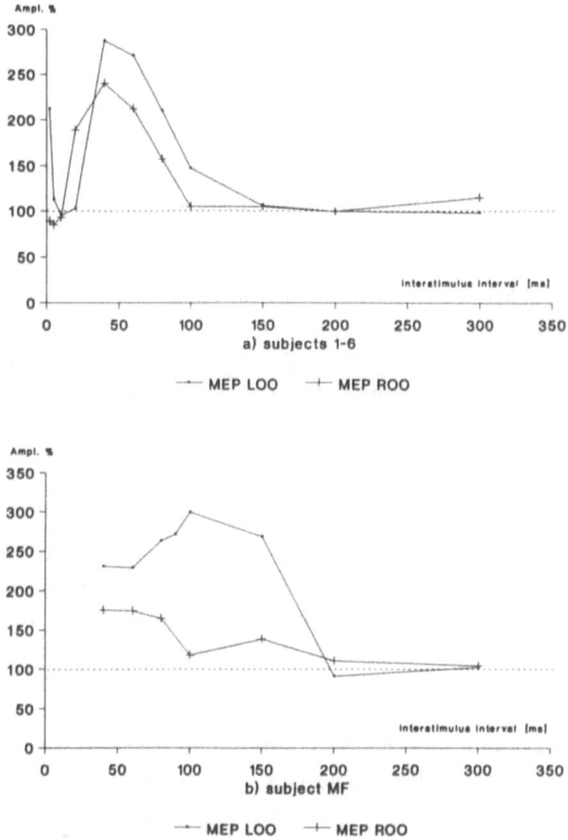

Fig. 2 a,b. Excitability curves of MEP amplitudes: peak-to-peak amplitudes at different interstimulus intervals (ISI) to a conditioning BR converted into percentages of the control response, target muscle at rest in each case (see text); *LOO*, left; *ROO*, right orbicularis oculi muscle. **a** Mean percentage values of six subjects; **b** percentage values (at least mean of two amplitudes per evaluated ISI between 40 and 300 ms only) of subject MF with constant R3 responses of the conditioning BR slight voluntary tonic innervation of OOM to facilitate the MEP [10]. Step 3 – paired stimulation (Fig. 1).

Evaluation: Recordings were stored on floppy discs for off-line analysis; amplitudes of the contralateral MEP measured peak-to-peak after conditioning BR were converted into percentages of the control response by single magnetic stimulation; heteronymous excitability curves are established (Fig. 2).

Results

As the supraorbital nerve extends to the parietal region, every magnetic stimulus delivered over the skull to the corresponding face area also elicits a BR via the trigeminofacial pathway by magnetic stimulation of the supraorbital nerve. For this, there was a double discharge of the contralateral OOM at single magnetic stimulation: the MEP at a latency of about 10 ms was followed by a late BR response, R2xm (Fig. 1). By paired stimulation this late response was significantly inhibited, up to total depression (Fig. 1). This inhibition was observed at ISI of 40–300 ms in our study; at shorter ISI evaluation of R2xm was impeded by foregoing late responses of the conditioning BR. Irrespective of this inhibition, the MEP was facilitated up to some 300% only at ISI, resulting in an approximately synchronous recording of the (test) MEP response with the (conditioning) BR response. Recording of the MEP between the activity of R1 and R2 on the left, before R2 on the right side, and after the late responses on both sides showed neither facilitation nor significant inhibition (Fig. 2).

Discussion

Corticofacial magnetically evoked potentials (MEP) of the OOM conditioned by electrical BR are facilitated, depending on the activity of the BR responses R1, R2(x), and R3(x). This is shown well in subject MF, who had constant R3 responses; consequently, his excitability curve was extended to longer ISI than the mean excitability curve of the other six subjects (Fig. 2b). Irrespective of this facilitation, the pathways of the late BR responses were inhibited by the conditioning BR; this was demonstrated by depression of the late BR responses, unavoidably evoked by simultaneous magnetic stimulation of the supraorbital nerve. This inhibition is due to reduced excitability at the interneuronal level of the late BR responses [6, 13], according to the known course of the excitability curve of the BR.

Facilitation of MEP was therefore observed, if there was nearly a synchronous arrival of impulses of both the trigemino- and corticofacial afferents at the common efferent motor neurons; this facilitation was shown to be independent of simultaneously inhibited interneurons of the late BR responses. Hence, facilitation of MEP by conditioning BR may be due to postsynaptic potentiation at motor neurons of the FN.

These results allow no unequivocal answer to our question about a functional linkage of cortico- and trigeminofacial interneurons. Voluntary activation of the OOM has an excitatory influence on habituation of BR interneurons [2]. Comparably, the corticofacial pathways activated by magnetic stimulation may overcome the inhibition of the interneurons assumed as a common link. This may be all the more true if there are direct and indirect corticofacial connections in parallel and if relatively strong stimuli are used, as in our experimental setting. Another probable reason for missing inhibition of the MEP despite inhibited late BR responses might be the existence of independent cortico- and trigeminofacial interneurons, perhaps depending on the different – namely motor and sensory – afferents to the FN [4]. However, the crucial point is that of transferability of results obtained in the monkey to man [5]. Nevertheless, our results obtained electrophysiologically are at least consistent with the assumption that there are indeed only direct (crossed) corticofacial connections to the contralateral motor neurons of OOM in man, in accord with anatomical observations [8].

One may speculate about this as a specific difference from the results obtained in the monkey [5], thereby pointing at the difference in species, such as with facial expression in man. This latter is a further topic, beyond the primary range of our study. However, the excitability curves give a possible hint at this: the facilitation of MEP is pronounced in the *left* OOM (Fig. 2). Further investigation is necessary as to whether this side-to-side difference is due to left supraorbital stimulation. However, we want to stress the resemblance to two additional observations: first, the imbalanced suprapontine influences upon the efferent BR centers in the brain stem resulting in a predominance of the *left* in most subjects [7], and second, the extensively discussed asymmetrical facial expression in man with *left* facial dominance [12].

Acknowledgement. We thank Mrs. D. Dokk for her excellent technical assistance.

References

1. Benecke R, Meyer BU, Schönle P, Conrad B (1988) Transcranial magnetic stimulation of the human brain: responses in muscles supplied by cranial nerves. Exp Brain Res 71:623–632
2. Desmedt JE, Godaux E (1976) Habituation of exteroceptive suppression and of exteroceptive reflexes in man as influenced by voluntary contraction. Brain Res 106:21–29
3. Holstege G, Kuypers HGJM, Dekker JJ (1977) The organization of the bulbar fibre connections to the trigeminal, facial and hypoglossal motor nuclei. II. An autoradiographic tracing study in cat. Brain 100:265–286

4. Holstege G, van Ham JJ, Tan J (1986) Afferent projections to the orbicularis oculi motoneuronal cell group. An autoradiographical tracing study in the cat. Brain Res 374:306–320
5. Jenny AB, Saper CB (1987) Organization of the facial nucleus and corticofacial projection in the monkey: a reconsideration of the upper motor neuron facial palsy. Neurology 37:930–939
6. Kimura J, Harada O (1976) Recovery cycle of the blink reflex during wakefulness and sleep. J Neurol 213:189–198
7. Kossev A, Dengler R, Struppler A (1983) Quantitative assessment of the blink reflex in normals. Physiological side-to-side differences and frequency dependence. Electromyogr Clin Neurophsiol 23:501–511
8. Kuypers HGJM (1958) Corticobulbar connexions to the pons and lower brain stem in man. Brain 81:364–388
9. Malin J-P (1982) The human orbicularis oculi reflex. Some principal remarks. Electromyogr Clin Neurophsiol 22:45–53
10. Meyer B-U, Britton TC, Benecke R (1990) Magnetic stimulation of the corticonuclear system and of proximal cranial nerves in humans. In: Berardelli A et al. (eds) Motor disturbances II. Academic, London, pp 235–248
11. Rimpel J, Geyer D, Hopf HC (1982) Changes in the blink responses to combined trigeminal, acoustic and visual repetitive stimulation, studied in the human subject. Electroencephalogr Clin Neurophysiol 54:552–560
12. Thompson JK (1985) Right brain, left brain; left face, right face: hemisphericity and the expression of facial emotion. Cortex 21:281–299
13. Willer JC, Roby A, Boulu P, Albe-Fessard D (1982) Depressive effect of high-frequency peripheral conditioning stimulation upon the nociceptive component of the human blink reflex. Lack of naloxone effect. Brain Res 239:322–326

Facial Myokymia in Brain-Stem Disorders

L. Gutmann

Department of Neurology, West Virginia University Health Sciences Center,
Morgantown, WV 26506, USA

Introduction

Myokymia may occur in numerous muscles innervated by cranial and spinal nerves, but it is seen most commonly in those innervated by the facial nerve. The presence of facial myokymia indicates a lesion involving the facial nerve, either within the pons or peripherally, and can be helpful in identifying the nature of the underlying problem.

Facial myokymia occurs with a variety of disorders, most notably the Guillain-Barré syndrome [5, 8, 20, 25, 34], multiple sclerosis [1, 14, 15, 24, 25, 35], brain-stem tumors [10, 18, 22, 25, 29, 35], and timber rattlesnake (*Crotalus horridus horridus*) envenomation [4, 6]. Schultze [30] used the term "myokymia" in 1894 in describing continuous undulating muscle movements in the legs. Bernhardt [2] first described facial myokymia in 1802, and Oppenheim [24] associated it with multiple sclerosis in 1916. The clinical phenomenon is characterized by undulating, rippling, and wavelike movements spreading across the face. It persists during sleep. When prominent,it may be associated with a continuous contraction of the orbicularis oris and orbicularis oculi muscles [23, 32]. The characteristic underlying electrophysiological abnormality is known as myokymic discharges. These are spontaneous tetanic bursts of individual motor unit potentials recurring rhythmically or semi-rhythmically, usually several times per second. Myokymia may be sufficiently gross to be easily appreciated by the patient and physician. On the other hand, it may be very subtle, requiring electrophysiological examination to verify its presence.

Physiological Considerations

The myokymic discharges are likely the result of a biochemical abnormality in the microenvironment of the motor axon membrane anywhere between the motor neuron and the terminal axon [5, 6, 13, 15]. This results in an increased level of membrane excitability. The exact location of this myokymic discharge generator varies with the underlying disorder.

L.R. Caplan and H.C. Hopf (Eds.)
Brain-Stem Localization and Function

The disorders that most commonly result in myokymic discharges appear to alter the biochemical microenvironment of the facial nerve axons by the process of demyelination, direct toxic effects, ischemia, hypoxia, and edema. Radiation changes, an uncommon cause of facial myokymia, is an important one of myokymia in the extremities [9, 16, 19, 33]. Axon compression is an uncommon cause of myokymic discharges [1, 31] and pontine infarctions, in particular, do not result in myokymia. The manner in which brain-stem tumors [10, 18, 22, 25, 29, 35] and other brain-stem masses produce myokymic discharges is not clear, especially since the myokymia may persist for many years in this circumstance. It is tempting to speculate that it may be due to the chemical changes related to edema at the intrapontine portion of the axon.

The clinically observable myokymia results from myokymic discharges arising in multiple motor units. Myokymic discharges are typically short bursts of ectopically generated motor unit potentials, with each burst containing potentials often firing at a rate of 20–100 Hz. Slower or faster rates may occur [1, 5, 20]. Myokymic bursts may occur as doublets (Fig. 1), triplets, or multiplets, some multiplets being sufficiently prolonged to persist for several seconds (Fig. 2) [13]. The bursts of myokymic discharges usually recur in a rhythmic or semi-rhythmic fashion. In general, short bursts (i.e., doublets) recur at a more rapid rate than more prolonged discharges (i.e., multiplets) (Figs. 1 and 2). Some prolonged multiplets may recur only several times per minute [13].

The existence of an ectopic generator as the cause of the myokymic discharges was first suggested by Rasminsky, based on his studies of demyelinated axons in dystrophic mice [26]. Manipulating the Ca^{++} in the microenvironment of the ectopic generator alters the excitability of the axon membrane, and thereby the firing characteristics of the myokymic discharges

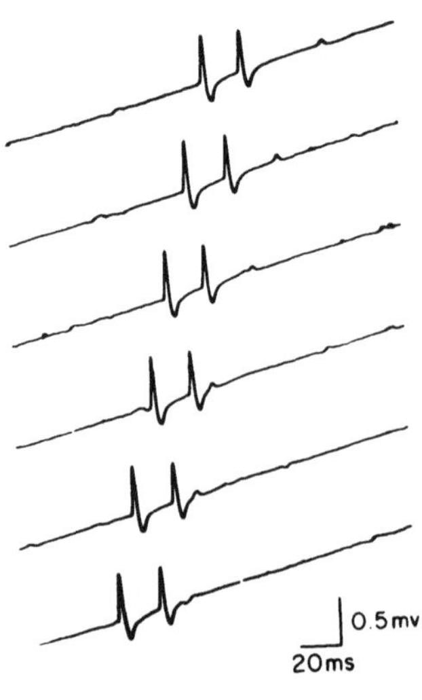

Fig. 1. A doublet recurring regularly from the orbicularis oris of a patient with Guillain-Barré syndrome. Continuous sweeps are recorded from *bottom* to *top*

[5, 6, 15]. In peripheral axons, decreasing the external Ca^{++} concentration lowers the threshold for excitation, while increasing the Ca^{++} concentration raises it. Tetany resulting from the hypocalcemia of hypoparathyroidism is associated with an increased motor axon excitability and the spontaneous firing of doublets, triplets, and multiplets identical to myokymic discharges [11].

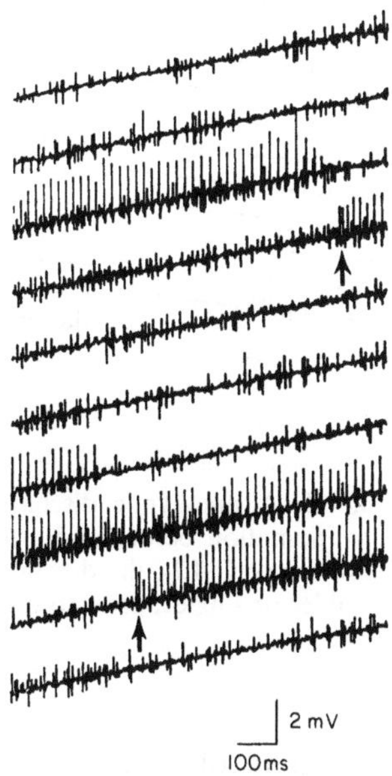

Fig. 2. A multiplet recorded twice (onset marked by *arrow*) from the orbicularis oris of a patient with a pontine glioma. Background activity reflects other myokymic discharges at some distance from the recording electrode. Continuous sweeps are recorded from *bottom* to *top*

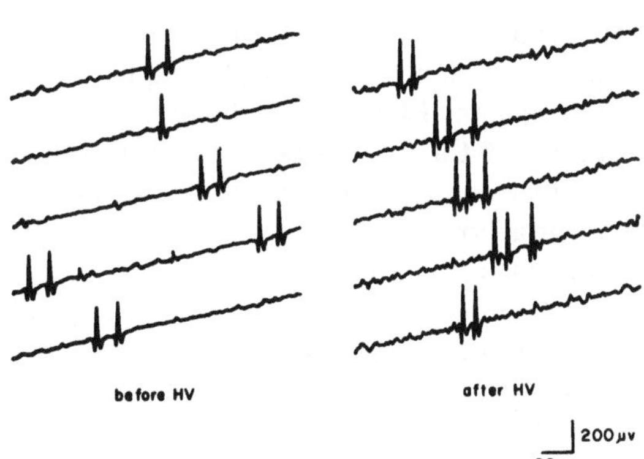

Fig. 3. A myokymic discharge recorded from orbicularis oris before and after 1 min of hyperventilation. Prior to hyperventilation, the discharge appeared as a doublet or singlet. Following hyperventilation, it appears as a triplet or doublet. Note that the interval between the first two spikes in each myokymic burst is shorter after hyperventilation than before. Continuous sweeps are recorded from *bottom* to *top*

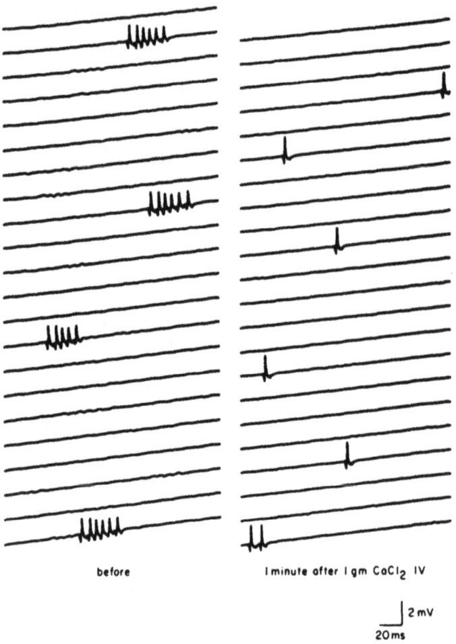

Fig. 4. A myokymic discharge was recorded from a patient with timber rattlesnake envenomation before and after the intravenous injection of 1 g $CaCl_2$. Prior to administration of $CaCl_2$, the discharge appears as a multiplet. One minute later, it appears as a doublet or singlet, prior to disappearing. Continuous sweeps are recorded from *bottom* to *top*

The myokymic discharges show burst amplification (increased number of motor unit potentials and shortening of the interspike interval within each burst) in the facial myokymia of the Guillain-Barré syndrome and timber rattlesnake envenomation when the serum ionized Ca^{++} is lowered with hyperventilation (Fig. 3) or plasmapheresis. The latter uses a Ca^{++}-binding anticoagulant. Elevation of the serum ionized Ca^{++} by intravenous administration of Ca^{++} chloride in these disorders results in prolongation of the interspike intervals within the myokymic burst, with subsequent fewer recurrent action potentials in each burst and, ultimately, disappearance of the recurrent burst firing (Fig. 4) [5, 6, 13]. The ectopic generator in each of these disorders is likely to be in the peripheral portion of the motor axon.

When the lesion involving the facial nerve occurs in the intrapontine portion of the axon, the response of myokymic discharge to changes in ionized Ca^{++} is absent or diminished (Figs. 5 and 6) [13, 15]. This has been documented in multiple sclerosis and pontine glioma (the latter, a recent unpublished observation). The reason for this absent or decreased response most likely reflects the blood brain barrier impeding acute fluxes of ionized Ca^{++} [28]. Biochemical differences in peripheral and central myelin might also play a role in this altered responsiveness.

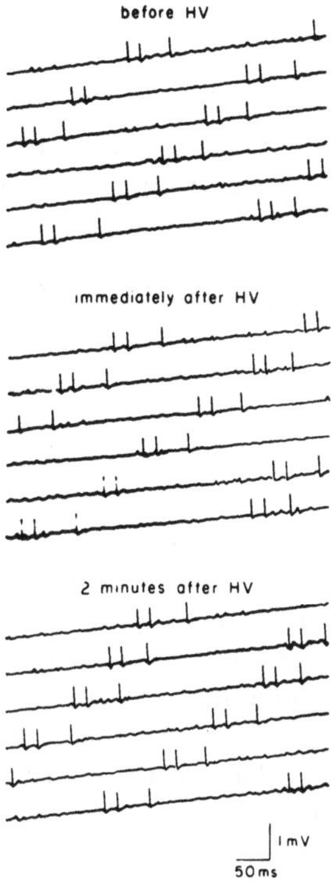

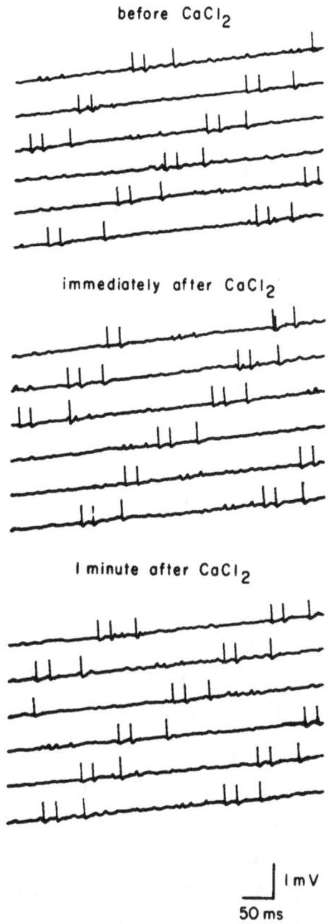

Fig. 5 (*left*). Myokymic discharge from the orbicularis oris of a patient with multiple sclerosis. The myokymic discharge appears as a triplet or doublet, which remains unaltered following hyperventilation for 2 min. Continuous sweeps are recorded from *bottom* to *top*

Fig. 6 (*right*). Myokymic discharge from the orbicularis oris of a patient with multiple sclerosis. The discharge appears as a triplet or doublet and is not altered by the intravenous infusion of 1 g $CaCl_2$. Continuous sweeps are recorded from *bottom* to *top*

Local nerve blocks result in a gradual disappearance of myokymic discharges if placed distal to the location of the ectopic generator [1]. Facial myokymic discharges due to timber rattlesnake envenomation result from the hematogenous spread of the venom from the bitten extremity. These discharges rapidly disappear within several hours after the administration of antivenin [6]. This observation provides strong support for the hypothesis that biochemical changes within the microenvironment of the axon result in the altered nerve excitability and the resultant myokymic discharges.

Clinical Manifestations

Facial myokymia is most often seen in the Guillain-Barré syndrome [4, 8, 20, 25, 34], multiple sclerosis [1, 14, 15, 24, 25, 35], and pontine tumors, most often gliomas [10, 18, 22, 25, 29, 35]. In sharp contrast, it is not seen with brain-stem infarctions, although it occasionally occurs with brain-stem ischemia following cardiopulmonary arrest [21]. It occurs invariably in the face following timber rattlesnake envenomation but disappears rapidly following the administration of antivenin [5, 6]. Facial myokymia occurs occasionally with metastatic, cerebellar, and cerebellopontine angle tumors [10, 25], pontine tuberculoma [3], meningeal carcinomatosis sarcomatosis [25], syringobulbia [25], lymphocytic meningoradiculitis [12], basilar invagination [7], and radiation [25].

Facial myokymia and myokymic discharges occur in 15% of patients with the Guillain-Barré syndrome [20]. This may be sufficiently obvious that the patient complains of it. The discharges may be so plentiful that a mild persistent contraction of the orbicularis oculi and orbicularis oris occurs. In other patients, the facial myokymia may be subtle or imperceptible requiring electrophysiological recordings to document its presence. The facial myokymic discharges in the Guillain-Barré syndrome are frequently bilateral and often associated with at least mild facial weakness. They appear within the first few days of the illness and persist up to 1 month [20].

In multiple sclerosis, the myokymia and myokymic discharges in the face are usually unilateral and transient. They are generally present for only a few weeks but may persist for up to 3 months. Episodes occasionally recur and may be either on the same or opposite sides [14, 15]. Facial weakness may be present, as well as a continuous contraction of the orbicularis oculi and orbicularis oris muscles when the myokymia is especially pronounced [9, 23, 32].

Facial myokymia in the presence of a posterior fossa tumor or mass occurs occasionally. Pontine glioma is a most likely posterior fossa mass lesion with extramedullary masses producing myokymia only very occasionally. In addition to pontine gliomas, facial myokymia has been described with cerebellar astrocytomas, metastatic tumors, and acoustic neuromas. The myokymia in this circumstance is unilateral and, in contrast to the Guillain-Barré syndrome, multiple sclerosis, and timber rattlesnake envenomation, is usually persistent [10, 18, 22, 25, 29, 35]. Autopsy studies show edema involving the area of the facial nucleus with the tumor sparing the nucleus [22, 29, 33, 35]. A recent patient with pontine glioma had persistent myokymia and persistent facial muscle contraction for 17 years (unpublished observation). Facial myokymia is seen only rarely in muscles innervated by other cranial nerves in the presence of a brain-stem tumor [18].

The facial myokymia invariably present with timber rattlesnake envenomation is a remote effect due to the hematogenous spread of the toxin. It is always bilateral and disappears within several hours after the patient receives antivenin. It persists in the bitten hand or leg for an additional 3 days before clearing [6].

Therapeutic Considerations

Myokymia is usually asymptomatic and requires no therapy. In the occasional patient in whom the myokymia is sufficiently severe to produce a persistent contraction of the facial muscles, phenytoin or carbamazepine in anticonvulsant dosages may be successful in reducing the number of myokymic discharges [5, 17, 27].

References

1. Albers JW, Allen AA, Bastron JA, Daube JR (1981) Limb myokymia. Muscle Nerve 4:494–504
2. Bernhardt M (1902) Ein ungewöhnlicher Fall von Fazialiskrampf. Neurol Zentralbl 21:689–695
3. Boghen D, Filiatrault R, Descarries L (1977) Myokymia and facial contracture in brain stem tuberculoma. Neurology (Minncap) 27:270–272
4. Brick JF, Gutmann L (1988) Abduction-adduction hand tremor of myokymia. Neurology 38:995
5. Brick JF, Gutmann L, McComas CF (1982) Calcium effect on generation and amplification of myokymic discharges. Neurology (NY) 32:618–622
6. Brick JF, Gutmann L, Brick J, Apelgren KN, Riggs, JE (1987) Timber rattlesnake venom-induced myokymia: evidence of peripheral nerve origin. Neurology 37:1545–1546
7. Chalk CH, Litchy WJ, Ebersold MJ, Kispert DB (1988) Facial myokymia and unilateral basilar invagination. Neurology 38:1811–1812
8. Daube JR, Kelly JJ, Martin RA (1979) Facial myokymia with polyradiculoneuropathy. Neurology 29:662–669
9. DeSilva KL, Pearce J (1972) Facial myokymia: a clue to the diagnosis of multiple sclerosis. Postgrad Med J 48:657–661
10. Espinoza RE, Lambert EH, Klass DW (1967) Facial myokymia affecting the electroencephalogram. Mayo Clin Proc 42:258–270
11. Feldman RG, Granger CV (1963) Electromyography in latent tetany. N Engl J Med 269:1064–1067
12. Gemignani F, Juvarra G, Calzetti S (1981) Facial myokymia in the course of lymphocytic meningoradiculitis: case report. Neurology (NY) 31:117–118
13. Gutmann L (1991) Facial and limb myokymia. Muscle Nerve 14:1043–1049
14. Gutmann L, Thompson HG, Martin JD (1969) Transient facial myokymia. JAMA 209:389–390
15. Gutmann L, Brick JF, Riggs JE (1986) Calcium and myokymia of brainstem origin. Neurology 36:126–130
16. Harper CM, Thomas JE, Cascino TL, Litchy WJ (1989) Distinction between neoplastic and radiation-induced brachial plexopathy, with emphasis on the role of EMG. Neurology 39:502–506
17. Hughes RC, Matthews WB (1969) Pseudomyotonia and myokymia. J Neurol Neurosurg Psychiatry 32:11–14
18. Kaeser HE, Skorpil V (1976) Myokymia involving muscles innervated by the Vth, VIth VIIth, IXth, Xth, XIIth cranial nerves with brain stem tumor. Eur Neurol 14:408–412

19. Levin KH, Lederman RJ, Wilborn AJ (1987) Spectrum of EMG changes in radiation brachial plexopathy. Muscle Nerve 10:656
20. Mateer JE, Gutmann L, McComas CF (1983) Myokymia in Guillain-Barré syndrome. Neurology (NY) 33:374–376
21. Morris HH, Estes ML (1981) Bilateral facial myokymia following cardiopulmonary arrest. Arch Neurol 38:393–394
22. Negri S, Caraceni T, DeLorenzi L (1976) Facial myokymia and brain stem tumor. Eur Neurol 14:108–118
23. O'Connor, Wynn Parry CB, Davis R (1966) Continuous muscle spasm in intramedullary tumors of the neuraxis. J Neurol Neurosurg Psychiatry 29:310–314
24. Oppenheim H (1916) Neue Beitrage zur Klinik der multiplen Sklerose. Jahresber Neurol Psychiatr 20:14–18
25. Radu EW, Skorpil V, Kaeser HE (1975) Facial myokymia. Eur Neurol 13:499–512
26. Rasminsky M (1980) Ephaptic transmission between single nerve fibers in the spinal nerve roots of dystrophic mice. J Physiol (Lond) 305:151–169
27. Schaaf M, Payne C (1966) Effect of diphenylhydantoin and phenobarbital on overt and latent tetany. N Engl J Med 274:1228–1233
28. Schain RJ (1964) Cerebrospinal fluid and serum cation levels. Arch Neurol 11: 330–333
29. Sethi PK, Smith BH, Kalyanaraman K (1974) Facial myokymia: a clinicopathological study. J Neurol Neurosurg Psychiatry 37:745–749
30. Schultze F (1894) Beiträge zur Muskelpathologie. Dtsch Z Nervenheilkd 6:65–75
31. Streib EW (1990) Distal ulnar neuropathy as a cause of finger tremor: a case report. Neurology 40:153–154
32. Tenser RB, Corbett J (1974) Myokymia and facial contraction in brain stem glioma. Arch Neurol 30:425–427
33. Thomas JE, Cascino TL, Earle JD (1985) Differential diagnosis between radiation and tumor plexopathy of the pelvis. Neurology (NY) 35:1–7
34. Wasserstrom WR, Starr A (1977) Facial myokymia in the Guillain-Barré syndrome. Arch Neurol 34:576–577
35. Waybright EA, Gutmann L, Chou SM (1979) Facial myokymia pathological features. Arch Neurol 36:244–245

Persistent Tonic Facial Contraction: A Local Brain-Stem Sign

H.C. Hopf

Neurologische Klinik, Universität Mainz, Langenbeckstraße 1, 55131 Mainz, Germany

Impairment of facial motility associated with contralateral hemiparesis is characteristic of pontine lesions. In the Millard-Gubler and Foville syndromes, involvement of the facial nerve is nuclear or infranuclear, resulting in a peripheral-type palsy. Conditions of pontine origin showing hyperactivity of facial muscles are facial myokymia and Brissaud's syndrome. Brissaud and Sicard [4] reported low-frequency phasic cramplike facial contractions and contralateral hemiparesis in inflammatory brain-stem processes.

Another phenomenon, presenting as tonic facial contraction persisting over weeks and even months, has attracted my attention during the past several years. It has been observed associated with myokymia [13] but may occur separately [2, 19].

Clinical Picture

Clinically, persistent tonic facial contraction (PTFC) is easily recognized. The angle of the mouth is drawn outward and upward, the nasolabial fold is deepened, the palpebral fissure is narrowed, and the eyebrow is pulled downward, giving the face the appearence of a unilateral grin or contralateral facial palsy. The contraction may be strong enough to close the eye [15, 20, 21].

Optional associated signs are impaired motility of the involved muscles on volition, contralateral hemiparesis, myokymia, trigeminal sensory impairment, and blink reflex abnormalities indicating damage of the trigeminal afferent input.

Differentiation from Myokymia

Matthews [13] and Gutmann et al. [9] stated that persistent contraction is a feature of myokymia, although they did not mention it in three of their seven patients. In early individual reports both conditions have been observed coincidentally [3, 5, 14, 16]. Later descriptions, however, report that

L.R. Caplan and H.C. Hopf (Eds.)
Brain-Stem Localization and Function

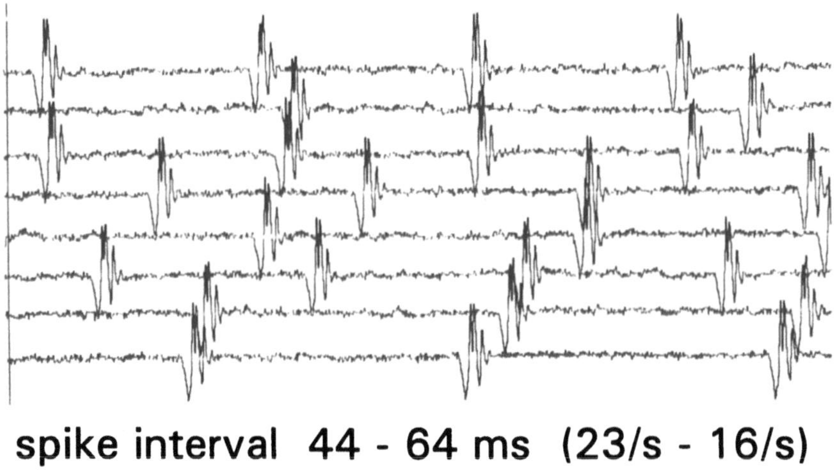

Fig. 1. Continuous motor unit discharges showing varying spike intervals in persistent tonic facial contraction (orbicularis oris muscle)

myokymia may occur without persistent contraction and vice versa: More than 50% of the patients with myokymia in the literature had no contraction [1, 6, 8, 9, 11, 13, 17]. Also, 50% of the patients with tonic contraction had no myokymia [18, 19].

Contraction in PTFC may be marked, even causing eye closure, which has not been reported with myokymia [2, 7, 12, 15, 20]. Repeatedly, trismus has been observed associated with PTFC but never with myokymia [18, 21]. Contralateral hemiparesis is a frequent sign with PTFC (68%) but is rarely seen in myokymia (14%).

The electromyographic pattern of PTFC consists of continuous motor unit activity. Spike intervals are 10–50 ms. Commonly, discharging of individual motor units is somewhat irregular and not affected by voluntary effort (Fig. 1). This pattern is completely different from myokymia, which shows the well-known doublets or multiplets with spike intervals of 4–50 ms within the bursts and burst intervals of 160–7000 ms [1, 8, 10].

Diagnostic Relevance

The coincidence of PTFC and trismus favors a mid to rostral pontine location. The frequent association with contralateral hemiparesis indicates the close anatomical relation to the pyramidal tract. In two patients MRI demonstrated a lesion in the base of the pons at the mid pontine level (Fig. 2).

PTFC occurs mainly with tumors (glioma) or ischemia of the brain stem. Myokymia is most frequently observed in multiple sclerosis and almost not seen in ischemia.

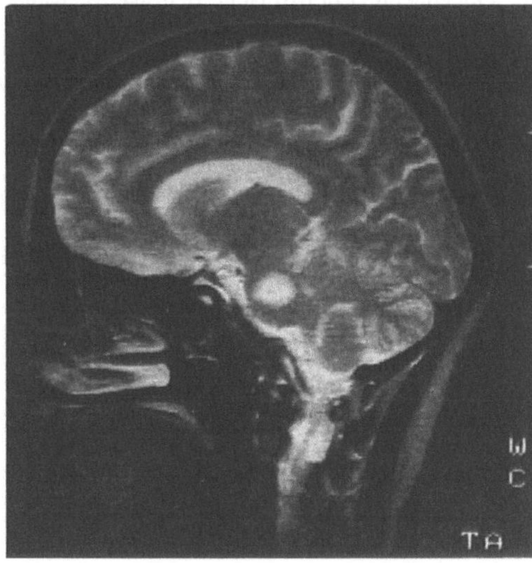

Fig. 2. MRI in persistent tonic facial contraction. Parasagittal section. TE 2315, TR 90, 0.5 Tesla

References

1. Andermann F, Cosgrove JBR, LLoyd-Smith DL, Gloor P, McNaughton FL (1961). Facial myokymia in multiple sclerosis. Brain 84:31–44
2. Arseni C, Petrovici I (1968). Persistent tonic facial spasm in brain stem tumours. J Neurol Sci 7:107–114
3. Bernhardt M (1902) Ein ungewöhnlicher Fall von Fazialiskrampf. Neurol Centralbl 21:689–695
4. Brissaud E, Sicard JA (1908) L'hémispasme facial alterne. Presse Med 16:234–241
5. Buzzard EF (1913) Varieties of facial spasm and their treatment. Practitioner 91: 745–754
6. DeSilva KL, Pearce J (1972) Facial myokymia: a clue to the diagnosis of multiple sclerosis. Postgrad Med J 48:657–661.
7. Draganesco S, Voiculesco V, Nicolesco SIT (1947) Spasme facial "fixe" au cours d'une tumeur intraprotuberantielle. Rev Neurol (Paris) 79:518–521
8. Espinosa RE, Lambert EH, Klass DW (1967) Facial myokymia. affecting the electroencephalogram. Mayo Clin Proc 42:258–270
9. Gutmann L, Thompson HG, Martin JD (1969) Transient facial myokymia. JAMA 209:389–391
10. Hjorth RJ, Willison RG (1973) The electromyogram in facial myokymia and hemifacial spasm. J Neurol Sci 20:117–126
11. Kino F (1928) Muskelwogen (Myokymie) als Frühsymptom der multiplen Sklerose. Dtsch Z Nervenheilkd 104:31–41
12. Kosary IZ, Shacked IJ, Ouaknine G, Braham J (1970) Continuous facial spasm with tumor of the pons. J Neurosurg 33:212–215
13. Matthews WB (1966) Facial myokymia. J Neurol Neurosurg Psychiatry 29:35–39
14. Newmark L (1903) Ein Fall von primärem tonischen Gesichtskrampf mit Muskelwogen. Neurol Centralbl 22:461–462
15. O'Connor PJ, Wynn Parry CB, Davies R (1966) Continuous muscle spasm in intramedullary tumors of the neuraxis. J Neurol Neurosurg Psychiatry 29:310–314

16. Oppenheim H (1916) Neue Beiträge zur Klinik der Multiplen Sklerose. Fortschr Ges amte Geb Neurol Psychiatry 20:14–18
17. Radü EW, Skorpil V, Kaeser HE (1975) Facial myokymia. Eur Neurol 13:499–512
18. Rasdolsky I 1935 Dorsal-pontines Tumorsyndrom. Z Ges amte Neurol Psychiatry 152:530–537
19. Sogg RL, Hoyt WF, Boldrey E (1963) Spastic paretic facial contracture. Neurology (Hinneap) 13:607–612
20. Tenser RB, Corbett JJ (1974) Myokymia and facial contraction in brain stem glioma. Arch Neurol 30:425–427
21. Thiebaut F, Isch F, Isch-Treussard C, Jesel M (1964) Contracture et myokymies faciales. Etude clinique et électromyographique des deux observations. Problème diagnostique. Rev Otoneuroophthalmol 36:136–143

One-and-a-half Syndrome and Facial Palsy of Peripheral Type: A Rare Brain-Stem Syndrome

J.R. Rauh, W. Obhof, W. Esser, and K.-F. Druschky

Neurologische Klinik, Städtisches Klinikum Karlsruhe, Moltkestraße 14, 76133 Karlsruhe, Germany

Introduction

In 1967 Fisher [4] described a characteristic disorder of horizontal eye movements in which, due to a lesion of the medial longitudinal fasciculus (MLF) of the pontomesencephalic tegmentum and of the ipsilateral pontine gaze center, there is an ipsiverisive conjugate horizontal gaze palsy and a contraversive inhibition of adduction of the ipsilateral eye – internuclear ophthalmoplegia (INO). Fisher called this the one-and-a-half syndrome.

The combination of a one-and-a-half syndrome and a peripheral-type facial palsy has been described in only a few individual patients to date and indicates a circumscribed lesion in the paramedian tegmentum of the pons.

Case Report

History

A 34-year-old man who had previously been healthy noticed blurred vision and, 2 days later, asymmetry of his face. He sensed a slight numbness around the right side of his mouth.

Neurologic Examination

There was a complete conjugate horizontal gaze palsy to the right side. The right eye could not be adducted past the midline, and there was a jerk nystagmus of the left eye on abduction. Vertical eye movements, convergence, and pupillary reactions were not affected, so that the criteria of the one-and-a-half syndrome were fulfilled. There was also a right facial palsy of the peripheral type, which involved equally all three branches. Bell's phenomenon was positive. There was no impairment of taste. No sensory deficit was detected, even in the perioral region. There was no paresis of the trunk and extremities. The tendon reflexes were normal on both sides.

L.R. Caplan and H.C. Hopf (Eds.)
Brain-Stem Localization and Function

There were no pathologic reflexes such as Babinski's sign. Coordination and the patient's mental state were normal.

Laboratory Studies

There were no signs of infection or inflammation. Serological screening for neurotropic organisms was negative. Cerebrospinal fluid gave a cell count and protein analysis that were normal. No oligoclonal bands were detected. Cytologically, lymphocytes were slightly elevated.

Electrodiagnostic Findings

The pattern-shift visual-evoked potentials and the somatosensory evoked potentials of the trigeminal nerve were normal. The brain-stem auditory evoked potentials were nonspecifically altered, with badly generated waves III and IV but normal latencies. In the electrically evoked orbicularis-oculi reflex all potentials of right-sided measurements were delayed and reduced in amplitude, correlating with the peripheral-type facial palsy.

Magnetic Resonance Imaging

Magnetic resonance imaging (Fig. 1) was done on a Tomicon (Bruker) with a supraconducting 1.5-Tesla magnet, using a slice thickness of 4 mm. At the level of the lower pons, in T_2-weighted (TR 2469, TE 75-200) and in proton-weighted (TR 2469, TE 25-50) images there was a hyperintense, well-delineated area, which did not show gadolineum enhancement. The area was localized in the paramedian tegmentum immediately ventral to the fourth ventricle, and it measured in the transverse plane 12 × 7 mm at maximum. It was demonstrable in three planes. In addition, there were small lesions in the left cerebellar hemisphere, the left putamen, and the rostral corpus callosum.

Course and Follow-up

Diagnostically, a multifocal encephalitis of unknown etiology was presumed. Following high-dose, initially intravenous, steroid therapy (500 mg prednisolone for 5 days) there was a quick and nearly complete recovery.

Three months later, there was only a slight weakness of the right frontal branch of the facial nerve. Cerebrospinal fluid was normal. Magnetic resonance imaging showed no change. Again 20 months later there was no change in the neurological state.

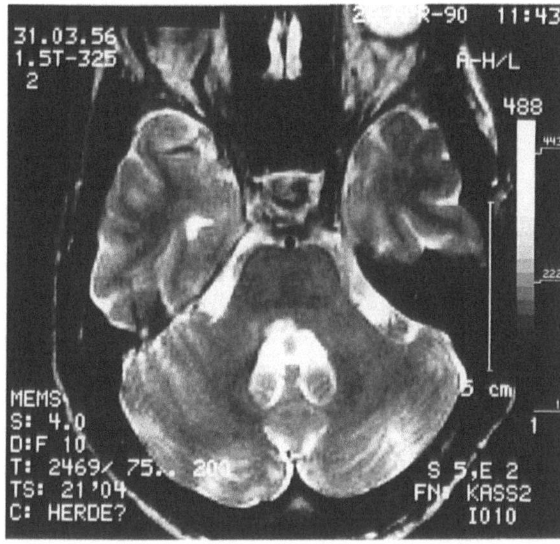

Fig. 1. Magnetic resonance imaging: T_2-weighted transverse plane at the lower pontine level, showing a well-delineated hyperintense area immediately ventral to the fourth ventricle

Discussion

Fisher's one-and-a-half syndrome is a relatively rare combination of symptoms. Fisher himself has reported on two women in the eighth decade of life who had the one-and-a-half syndrome. In the first case there was additional hemiparesis and dysarthria, and at necropsy an ipsilateral softening of the lower three quarters of the pons was found. In the second case there was additional severe dysarthria, dizziness, vomiting, unsteadiness on the feet, and a mild facial weakness on the side of the presumed lesion.

Wall and Wray [9] reviewed 29 cases of one-and-a-half syndrome from the literature as well as 20 of their own patients. They found, in addition to further oculomotor deficits such as upbeat nystagmus, skew deviation, and horizontal gaze nystagmus, combinations with other cranial nerve involvement. In four of their 20 patients an ipsilateral facial palsy of peripheral type was present. Other patients with one-and-a-half syndrome had, for example, abducens paresis, ipsilateral trigeminal sensory loss, and other neurological signs, such as incoordination or hemisyndrome. Etiologies were infarction, pontine hemorrhage, vascular malformation, neoplasm, and multiple sclerosis. Wall and Wray [9] pointed out that the one-and-a-half syndrome is caused by a lesion of the MLF and additionally a lesion of the ipsilateral paramedian pontine reticular formation (PPRF) and/or the ipsilateral abducens nucleus. Based only on the eye movement abnormality, it would not

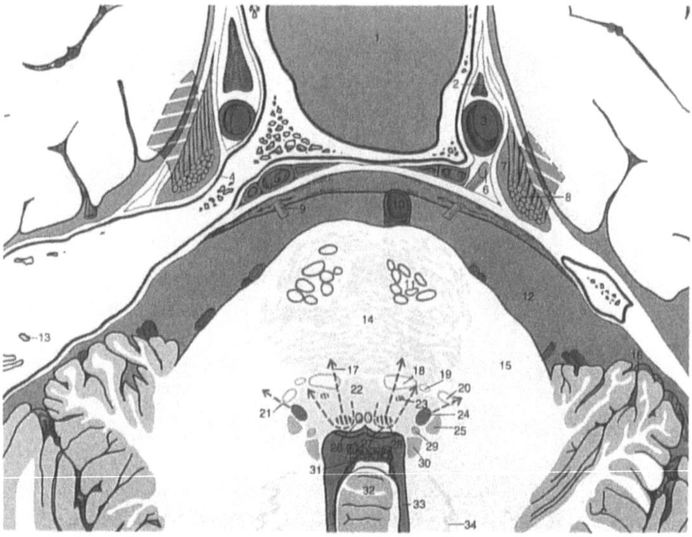

Fig. 2. Neuroanatomical correlation of the lesion shown in Fig. 1 in a schematic diagram of the brain stem in the same level. Nucleus abducens (*28*), genu nervi facialis (*27*), MLF (*26*) and parts of the PPRF (*22*) are involved. [From 6]

be possible to distinguish a lesion of the abducens nucleus from a lesion of the PPRF (see also [2]).

Hommel et al. [5] reported on a case of one-and-a-half syndrome with extensive further neurological disorders and widespread pontine lesions, demonstrated by MRI.

Balthasar [1] studied the rostrocandal localization of pontine lesions that cause abnormalities of horizontal gaze. He pointed out that the pathways of the PPRF as the presumed location of the premotor pontine horizontal gaze center and of the MLF run through the pons at all levels, and that therefore a horizontal gaze palsy or an INO alone cannot give a reliable topodiagnostic correlation. Among others, he reported on a patient in which there was the combination of a one-and-a-half syndrome of the right side, a facial palsy of central type on the left side, and a distinct hemisyndrome of the left side. At autopsy a small pea-sized metastasis was found in the middle level of the pons in the paramedian tegmentum. Balthasar defined this combination of symptoms as a syndrome of the intermedian paramedian tegmentum of the pons, as distinct from the Foville syndrome of the caudal pons.

In the case we have reported here there was a combination of a one-and-a-half syndrome and a right-sided peripheral-type facial palsy without impairment of taste and without other neurological signs. The neurological deficit was correlated with a very discrete focal lesion in the right caudal paramedian tegmentum of the pons. Figure 2 shows that the lesion demonstrated by MRI presumably involves the right abducens nucleus, parts of the right PPRF, the right MLF, and the genu of the facial nerve. There was no impairment of taste because the nervus intermedius, which contains the taste fibers of the anterior two thirds of the tongue, leaves the facial nerve after entering the brain stem and terminates in the nucleus solitarius [7].

The combination of the one-and-a-half syndrome with peripheral-type facial palsy has only rarely been reported, but this rare brain-stem syndrome is of high topodiagnostic value and indicates a focal lesion in the caudal paramedian tegmentum of the pons.

Acknowledgements. We wish to thank Dr. T. Niederstadt (*Zentrales Röntgeninstitut, Städtisches Klinikum Karlsruhe*, Direktor: Professor Dr. H.-J. Fernholz) for performing MR imaging and Dr. R. Murray for reviewing the manuscript.

References

1. Balthasar K (1973) Zur Lokalisation der horizontalen Blicklähmung und der internukleären Ophthalmoplegie. Arch Psychiatr Nervenkr 217:223–246
2. Brazis PW, Masdeu JC, Biller J (1990) Localization in clinical neurology. Little, Brown, Boston
3. Bronstein AM, Rudge P, Gresty MA, Du Boulay G, Morris J (1990) Abnormalities of horizontal gaze. Clinical, oculographic and magnetic resonance imaging findings. II. Gaze palsy and internuclear opthalmoplegia. J Neurol Neurosurg Psychiatry 53:200–207
4. Fisher CM (1967) Some neuro-ophthalmological observations. J Neurol Neurosurg Psychiatry 30:383–392
5. Hommel M, Gato JM, Pollack P, Borgel F, Reeves AG, Crouzet G, Lebas JF, Perret J (1987) Magnetic resonance imaging and the "one-and-a-half" syndrome. J Clin Neuro Ophth almol 7(3):161–164
6. Kretschmann HJ, Weinreich W (1991) Klinische Neuroanatomie und kranielle Bilddiagnostik. Thieme, Stuttgart
7. Nieuwenhuys R, Voogd J, van Huijzen C (1988) The human central nervous system, 3rd edn. Springer, Berlin Heidelberg New York
8. Oommen KJ, Smith MS, Labadie EL (1982) Pontine hemorrhage causing Fisher one-and-a-half syndrome with facial paralysis. J Clin Neuro Ophth almol 2:129–132
9. Wall M, Wray SH (1983) The one-and-a-half syndrome – a unilateral disorder of the pontine tegmentum: a study of 20 cases and review of the literature. Neurology (NY) 33:971–980

The Stapedial Reflex in Pontine Lesions

E. Lehnhardt

Hals-Nasen-Ohrenklinik, Medizinische Hochschule Hannover,
Konstanty-Gutschow-Straße 3, 30625 Hannover, Germany

The stapedial reflex (StapR) is elicited by excitation of the auditory nerve. The reflex loop runs across the cochlear nucleus and the superior olivary complex to the facial nuclei on both sides (Fig. 1). From there motor nerve fibers lead directly to the stapedius muscle. Its latency is approximately 10 ms [1, 5]. Using retrograde-transported horseradish-peroxidase tracers in guinea pigs, Strutz recently suggested (1992, personal communication) that the reflex pathway passes the lateral lemniscus as well. From a clinical point of view, this seems improbable. The contraction of the stapedius muscle in conjunction with the tensor tympani muscle prevents a nonlinearity of the ossicular chain to loud signals.

In routine audiological assessment, the StapR is activated with tonal impulses for 1.5 s. The change in middle ear impedance is registered over time. The most important measurement is that made of the intensity (dB) necessary to activate the reflex, e.g., the intensity at which a just noticeable change in the impedance is recorded. In the cases illustrated here, we have chosen not to provide the dB values and have only categorized as to positive (present, +) or negative (absent, ø) reflexes.

The results of contra- or ipsilateral measurements are optimally evaluated neuro-otologically if recorded in a specific manner [3, 4] (Fig. 2). To complete the diagnostic picture, the pure-tone audiogram must be considered together with tonal signals traveling through the various pathway stations of the brain stem, briefly discussed above. The stimulated ear is identified by means of a color mode: red for the right ear, blue for the left ear. The side of the form on which the impedance data are entered identifies whether the data have been obtained from the right or the left ear. Data on the left are for the right ear and data for the left ear are on the right. The color code is for the stimulated ear, and the location on the picture is for the measured ear.

Bilateral StapR abnormality following monaural stimulation (diagonal defect pattern, Fig. 3 ⟨1⟩) is due to lesions of the acoustic nerve or, if intra-axial, of the fibers to the facial nuclei close to their origin at the cochlear nucleus. Unilateral (left or right) StapR abnormality following stimulation of either ear (vertical defect pattern, Fig. 3 ⟨2⟩) is due to lesions of the

L.R. Caplan and H.C. Hopf (Eds.)
Brain-Stem Localization and Function

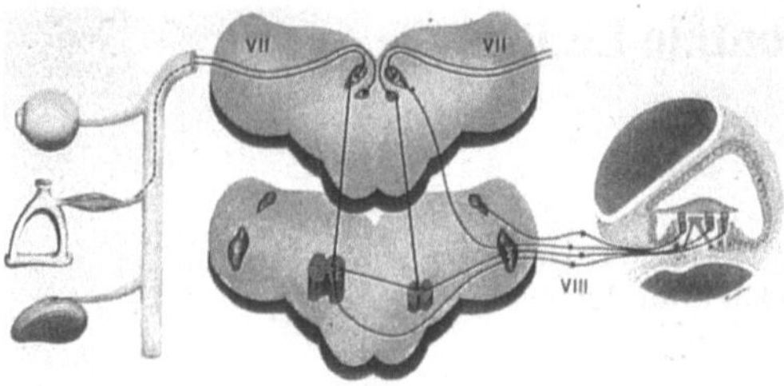

Fig. 1. The progression of the acoustic-facial stapedial reflex, half-schematic after Bosatra [2]

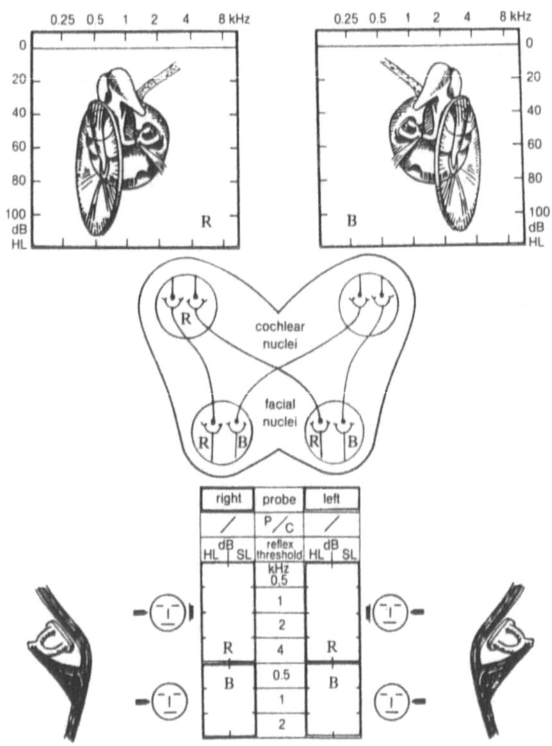

Fig. 2. Audiometric scheme for recording the stapedial reflex pattern, beginning by tone impulse at the ear drum, crossing the brain stem, and ending at the stapedius muscle. *R* means red, *B* means blue; for explanation see text.

facial nerve or around the facial nucleus involving the connections from both cochlear nuclei that converge on one facial nucleus. Contralateral StapR abnormality following stimulation of either ear (left and right) (medial defect pattern, Fig. 3 ⟨3⟩) is due to a lesion of only the crossing internuclear connections. Unilateral StapR abnormality ipsilateral to the stimulated ear

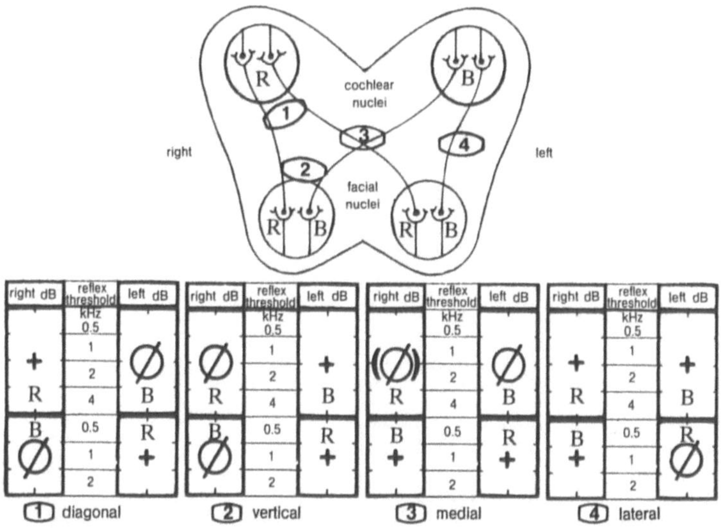

Fig. 3. Possible defect patterns of the stapedial reflex and the associated pathways and nuclei in the brain stem

(lateral defect pattern, Fig. 3 ⟨4⟩) is due to a lesion of only the uncrossed ipsilateral connections.

An example of a diagonal defect pattern is that of a woman (Fig. 4a) referred to us by a neurosurgeon who had preliminarily diagnosed a cerebellum tumor. Our objection was based on a pathological process occurring on the floor of the fourth ventricle, because we obtained a defect of the StapR by stimulating only the right ear. The surgeon, Prof. Brock graciously left us a drawing of the findings, actually showing a hematoma of the floor of the fourth ventricle to the right of the midline (Fig. 4b). A few days after removal of the hematoma the stapedial reflex was again intact, and on the right side only an auditory fatigue up to 50 dB remained. An identical defect pattern can also be shown with an acoustic neuroma or a small cerebello-pontine angle tumor, a common neurological finding.

It should be mentioned that after operation, a previously nonfunctional reflex can possibly become functional again. A patient, for example, even before surgery, can have normal audiometric results of the affected side, but the stapedial reflex can be absent. Then the return of the stapedial reflex after the operation should be interpreted more as being the result of pressure relief at the brain stem than as a recovery of the function of the auditory nerve. If a vestibular function is missing, then the diagonal defect pattern of the stapedial reflex suggests multiple sclerosis.

A vertical defect pattern normally is found in peripheral facial paresis or in abnormalities of the middle ear, which the otologist should be able to recognize or exclude. However, in the vertical irregular pattern, a finding in

a

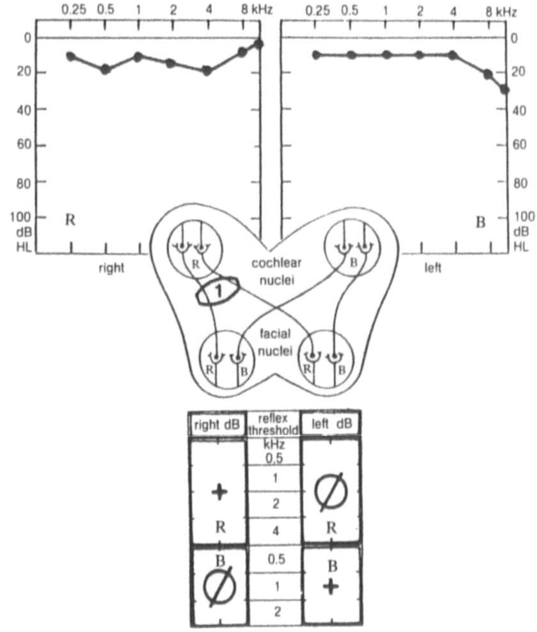

b

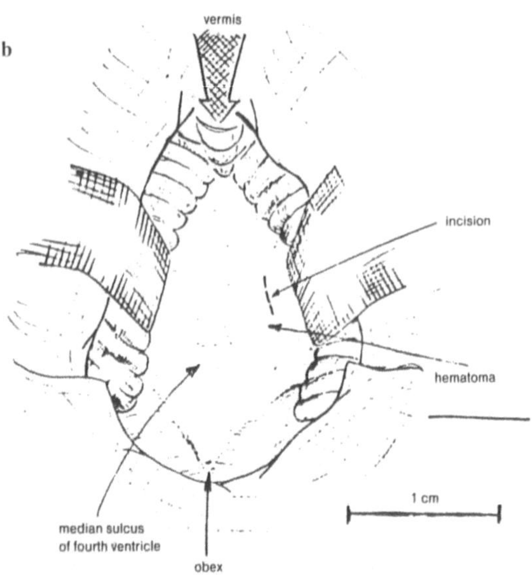

Fig. 4. a Diagonal irregular reflex pattern (*a*) in a hematoma on the floor of the fourth ventricle to the right of midline. **b** Operational site: hematoma of the floor of the fourth ventricle, right, next to the medial nucleus

the brain-stem may also be responsible, particularly when there is a lesion that blocks the fibers coming from the ipsi- or contralateral cochlear nuclei and is in the direct vicinity of the facial nuclei, for example in multiple sclerosis again (Fig. 2⟨2⟩).

We recorded a medial defect pattern from a man who had received a series of evaluations, and who had been diagnosed neurologically as having olivopontocerebellar atrophy. Consistent with this finding was a bilateral 50% reduction in dichotic speech understanding, despite approximately normal pure-tone hearing on both sides.

Only the crossed reflex pathway was affected in an 8-year-old boy with a medulloblastoma, while the laterally lying, ipsilateral connections were still functioning (Fig. 5). Isolated dysfunction of the crossed pathway due to ventral pressure can, albeit rarely, be limited to one side, as seen by us in a patient with an acoustic neuroma.

A strictly lateral process in or on the brain-stem indicates an abnormality which affects only the ipsilateral connection of the cochlear nucleus to the facial nucleus, as seen in a patient suffering from a meningioma of the tentorium. In addition to the isolated unilateral defect of the ipsilateral StapR, audiometric results confirmed such a lesion, along with a slight reduction of the dichotic speech understanding on the same side, which is also a brain-stem finding. The same defect pattern, however, is typical in the early stages of middle ear otosclerosis – a specific otologic disorder.

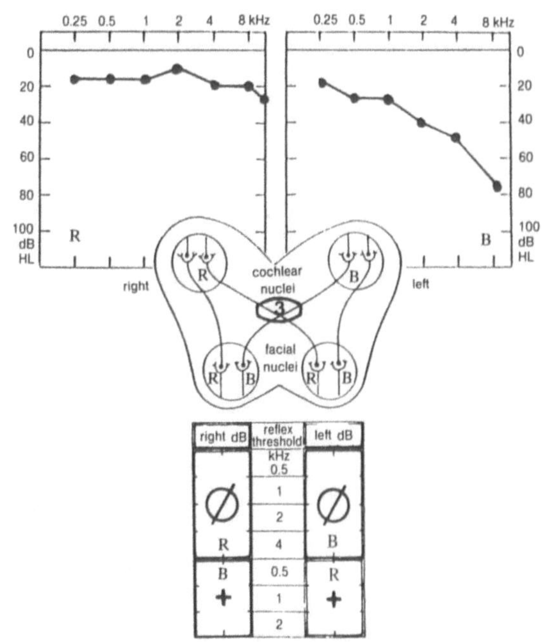

Fig. 5. Absence of crossed pathways through a medulloblastoma, medial site of lesion

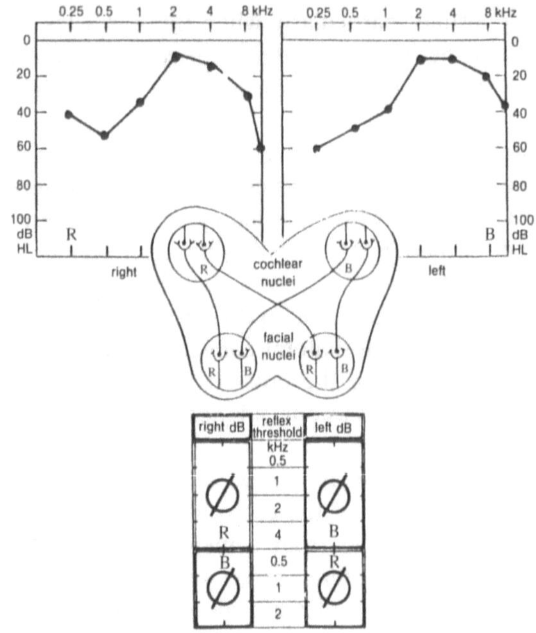

Fig. 6. Complete absence of the stapedial reflex in a 10-year-old boy with spinocerebellar degeneration

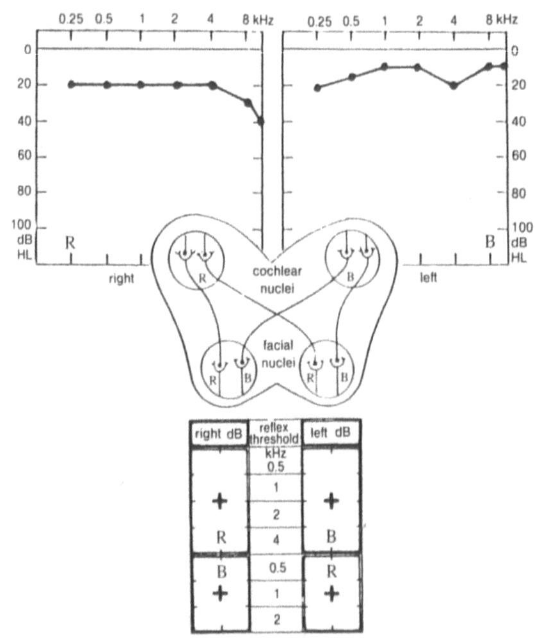

Fig. 7. Completely preserved stapedial reflex; tumor in the upper portion of the brainstem

With advanced otosclerosis, there is usually an absence of the impedance changes both contra- and ipsilaterally. The reason for this is the bony fixation of the stapes, which does not permit a movement of the stapes by contraction of the stapedius muscle. Such patients also have a conduction-type hearing loss.

Complete absence of the reflex could, finally, be caused by aplasia of the muscle, an extremely rare anomaly.

Complete absence of contra- and ipsilateral reflexes is illustrated by a 10-year-old boy with spinocerebellar degeneration (Fig. 6). This was explained neurologically. At the same time, speech understanding was considerably reduced and continued to decrease as the intensity of the stimuli increased, an apparent sign of neural or central origin of the malfunction.

If, on the other hand, there is a complete functional retention of the stapedial reflex because of a tumor in the region of the auditory pathway, then this constellation can possibly be used to localize the lesion beyond the reflex arc. However, in view of modern technologies, particularly magnetic resonance imaging, this finding is less important in the diagnostic process.

Figure 7 shows a patient who had auditory fatigue bilaterally, although the audiometric thresholds were only minimally elevated. Dichotic speech understanding was nevertheless completely absent on one side. CT scan showed the tumor to be located in the upper brain stem. For that reason, the stapedial reflex was able to function normally, despite the extreme auditory fatigue.

For the operating otologist, the stapedial reflex has recently attained a completely new meaning. In performing cochlear implants for deaf adults and deaf-born infants, we can, while still in the operating room, observe the function of the reflex using the operation microscope. As soon as the electrodes are placed near the auditory nerve, electrical stimuli of a specific strength cause contractions of the stapedius muscle. This contraction can be observed as a change of light reflection. From this we know that the reflex arc, even if it has never functioned before in deaf-born children, is functioning now. The electrical threshold values obtained during this process are a valuable reference for the programming of speech processors in children during rehabilitation later.

The smallest muscle in the human body, the stapedius is therefore important for neurology and for otology, not only in diagnostics but also for therapeutic applications.

References

1. Borg E (1973) On the neuronal organization of the acoustic middle ear reflex. A physiological and anatomical study. Brain Res 49:101–123
2. Bosatra A (1977) Pathology of the nervous arc of the acoustic reflex. Audiology 16:307–315

3. Jerger S, Jerger J (1975) Extra- and intra-axial brain stem auditory disorders. Audiology 14:93–117
4. Lehnhardt E (1983) Audiometrische Diagnostik im Hirnstamm. Laryngol Rhinol Otol (Stuttg) 52:11–21
5. Møller A (1062) Acoustic reflex in man. J Acoust Soc Am 34:1524–1534

Taste Distrubance Associated with Paramedian Posterior Thalamo-Subthalamic Artery Syndrome

B. Guldin and W. Christe

Abteilung für Neurologie, Universitätsklinikum Rudolf Virchow,
Freie Universität Berlin, Augustenburger Platz 1, 13353 Berlin, Germany

Introduction

Ischemic infarction in the territory of the paramedian posterior thalamo-subthalamic artery causes a well-known syndrome of oculomotor impairments consisting of down-gaze palsy, slowing of vertical saccades, and loss of convergence movements. Further symptoms, especially rostral mesencephalic oculomotor signs, may accompany the syndrome, e.g., upgaze palsy, impairment of the vertical vestibulo-ocular reflex, skew deviation or complete ocular tilt reaction, convergence retraction nystagmus, and pupillary abnormalities. This paper emphasizes the presence of a unilateral taste disturbance occurring simultaneously with these oculomotor findings.

The route of the gustatory system in man has not been identified as yet. The primary gustatory afferents project to the nucleus of the solitary tract. The rostral part of this nucleus contains neurons responding to chemicals which are effective taste stimuli [18]. There are only a few case reports of taste disturbances in human subjects following ipsilateral lesions in the nucleus of the solitary tract or its afferents with verification of these lesions on postmortem examination [14, 17]. Currently, it is assumed that the rostral half of the nucleus of the solitary tract is the first central gustatory relay [16]. The further route of gustatory information is still under discussion. Some investigators showed a crossed lemniscal projection from the nucleus of the solitary tract to the contralateral medial basal complex in guinea pigs [3], rabbits [9], and in one monkey [12]. On the other hand, there is evidence for an uncrossed ipsilateral projection of gustatory afferents via the central tegmental bundle to the ipsilateral ventral posteromedial thalamic nucleus. Benjamin [5] and Emmers [8] detected only ipsilateral receptive fields when recording from single units in the ventroposteromedial thalamic nucleus in cats and monkeys. Beckstaedt et al. [4] injected tritiated amino acids into the rostral part of the solitary tract in an electrophysiologically controlled experiment in an old-world monkey (cynomolgus monkeys) and were able to trace the projection from the rostral half of the nucleus of the solitary tract via the central tegmental bundle to the ipsilateral parvicellular part of the ventral posteromedial thalamic nucleus. There are only a

L.R. Caplan and H.C. Hopf (Eds.)
Brain-Stem Localization and Function

few reports on taste disturbances in human subjects following lesions rostral to the nucleus of the solitary tract, and these are inconsistent, reporting contralateral deficits after hemispheric [6] or a thalamic lesion [1]. Others have described ipsilateral deficits after a pontine hemorrhage [15] or after cerebral infarctions and tumors involving the anterior part of the insula and the operculum [13].

Case Report

After an unusually long night's sleep and difficulty awakening a 40-year-old woman noticed diagonal double vision and gait deviation to the left. Three hours later, she developed perioral numbness on the right side including the lips and tongue and taste disturbance on the right side of the tongue. On neurological examination 24 h later she presented with an up- and downgaze palsy for saccadic, pursuit, and vestibular eye movements, skew deviation (the right eye being lowest), head tilt to the left, and lateropulsion to the left. Upon fixation, convergence retraction nystagmus occurred. Moving her eyes horizontally, she had pseudo-abducens paresis in both eyes caused by the increased convergence tone. Her right pupil was mydriatic and the reaction to light and vergence was only sluggish; the right eyelid was retracted. She had a taste deficit on the right side of the tongue, characterized by an inability to recognize solutions of 0.55 mol glucose, 18.9 mmol quinine, 0.431 mol sodium chloride, and 0.357 mol citric acid, she had no difficulties tasting with the left side of her tongue. The perioral numbness had already disappeared. Fundus photography was done 5 days later and did not reveal a

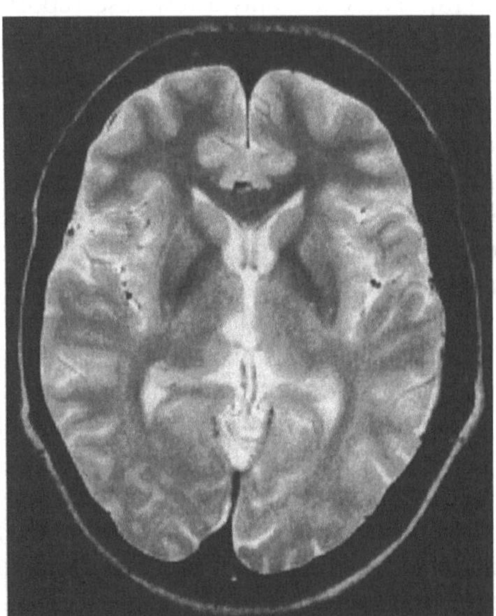

Fig. 1. Horizontal MRI demonstrating a subthalamic hyperintense lesion located paramedian on the right side of the mesencephalon (precontrast, T_2-weighted, TE 70, TR 2500)

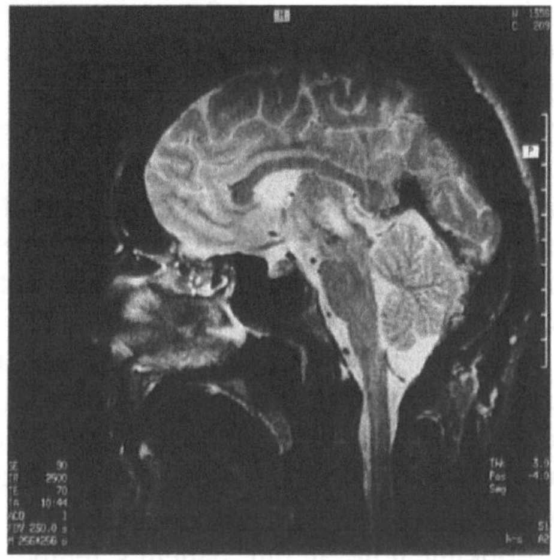

Fig. 2. Sagittal MRI showing the right-sided paramedian subthalamic hyperintense lesion (precontrast, T_2-weighted, TE 70, TR 2500)

cyclotropia of either eye. A week later, the MRI scan showed a single paramedian lesion in the rostral mesencephalon adjacent to the thalamus on the right side (see Figs. 1 and 2). The cerebrospinal fluid examination was normal; in particular, there was no evidence of an inflammatory process.

Discussion

The interesting findings in this case are the localization of the taste disturbance, namely ipsilateral to the side of the lesion, the ipsilateral perioral numbness, and the laterality of the skew deviation and head tilt: skew deviation to the right, head tilt and lateropulsion to the left, following a right-sided mesencephalic lesion.

We regard our case as supporting the hypothesis derived from the work of Beckstaedt et al. [4] that the gustatory system projects ipsilaterally in primates, including man. The findings of Beckstaedt and co-workers in 19 cynomolgus monkeys are at odds with the report of Kinney [12], who showed a crossed lemniscal pathway for the gustatory afferents in an old-world monkey (*Macaca mulatta*). However, Kinney [12] had four lesions in his preparation of a single monkey – two dorsal pontine and two medullary – so that it was rather difficult to trace the fine unmyelinated fibers from the nucleus of the solitary tract. Moreover, other fiber systems, especially from the trigeminal nuclei, could have interfered easily with the gustatory projection in his analysis. The reports on human cases with contralateral gustatory deficits caused by a hemispheric lesion were reviewed by Bornstein [6]. He did not include a case of ipsilateral gustatory impairment, because, from his point of view, an ipsilateral deficit could not be explained by a hemispheric lesion. The remaining 12 cases had lesions at the base of the pre- and postcentral gyrus and the underlying operculum. All of these patients had gustatory deficits more prominent contralateral to the lesion, but all of them also had somatosensory impairments contralaterally. These somatosensory deficits could have had an impact on gustatory testing. The third author suggesting a contralateral organization of the gustatory system was Adler [1], who described a patient with a glioblastoma invading the medial thalamus who had a contralateral taste deficit. Another patient, 2 years later [2], had a tumor invading the temporal lobe, again giving rise to contralateral gustatory impairment without olfactory or somatosensory deficits.

In contrast, among a series of 79 patients with cerebral lesions from tumors or infarctions who had been tested for gustatory impairment [13], none of them had a contralateral taste deficit but seven showed an ipsilateral taste deficit even though their motor deficits were on the contralateral side, as expected.

Looking now at our patient's temporary ipsilateral perioral numbness following a subthalamic or almost thalamic lesion, one has to take into account the experimental findings in an old-world monkey (*Macaca fascicularis*) made by Jones et al. [11], showing anatomically and physiologically that the medial 40% of the ventral posteromedial thalamic nucleus are devoted to somatosensory representation of the ipsilateral intra- and perioral structures. Given the close anatomic relationship between the somatosensory relay in the ventral posteromedial thalamic nucleus and the parvicellular part of the ventral posteromedial nucleus, their similar functional role as relay for the different sensory modalities of the oral cavity, and their paramedian localization, one might speculate whether the different sensory modalities are represented together ipsilaterally or bilaterally and possibly to a different extent ipsi-, contra-, or bilaterally across species and individuals.

The third puzzling finding in this case is a skew deviation to the right accompanied by head tilt and lateropulsion to the left after a right-sided mesencephalic lesion. According to Brandt [7], the ocular tilt reaction might be observed in patients with lesions along the projection route of the otolith and vestibular system. The ocular tilt reaction consists of skew deviation, head tilt to the side of the undermost eye, excyclotropia of the undermost eye, and incyclotropia of the other eye. Lateropulsion to the side of the undermost eye as well as a tilt of the subjective visual vertical are not part of the ocular tilt reaction but do evolve from lesions in the same projections; therefore, these symptoms may accompany ocular tilt reactions. A peripheral vestibular or pontine lesion would lead to an ipsiversive ocular tilt reaction, whereas a mesencephalic lesion (rostral to the crossing of the medial longitudinal fascicle) should lead to a contraversive ocular tilt reaction. In our patient the lateropulsion and head tilt are contraversive, but the skew deviation occurred in the opposite direction, i.e., ipsiversive to the lesioned side. To our knowledge, the only possible explanation for this finding is a monocular elevator palsy on the right side, obscuring the skew deviation. This explanation was first presented by Halmagyi et al. [10] in a review of patients with ocular tilt reaction. The course of oculomotor deficits in our patient favors this hypothesis: the vertical gaze palsy has recovered almost completely on the left eye but on upgaze the right eye stays behind and double vision occurs again.

References

1. Adler A (1933) Zur Topik des Verlaufs der Geschmackssinnsfasern und anderer afferenter Bahnen im Thalamus. Z Gesamte Neurol Psychiatr 149:208–220
2. Adler A (1935) Zur Topik der corticalen Geschmackssphäre. Z Gesamte Neurol Psychiatr 152:25–33
3. Allen WF, (1923) Origin and destination of the secondary visceral fibers in the guinea pig. J Comp Neurol 35:347–368
4. Beckstaed RM Morse JR, Norgren R (1980) The nucleus of the solitary tract in the monkey: projections to the thalamus and brain stem nuclei. J Comp Neurol 190: 259–282
5. Benjamin RM (1963) Absence of deficits in taste discrimination following cortical lesions as a function of the amount of preoperative practice. J Comp Physiol Psychol 52:255–258
6. Bornstein WS (1940) Cortical representation of taste in man and monkey. II. The localization of the cortical taste area in man, a method of measuring impairment of taste. Yale J Biol Med 13:383–387
7. Brandt T (1991) Vertigo. Springer, London
8. Emmers R (1966) Separate relays of tactile, pressure, thermal, and gustatory modalities in cat thalamus. Proc Soc Exp Biol Med 121:527–531
9. Gerebtzoff MA (1939) Les voies centrales de la sensibilité et du gout et leurs terminaisons thalamiques. Cellule 48:91–146
10. Halmagyi GM, Brandt T, Dieterich M, Curthoys IS, Stark RJ, Hoyt F (1990) Tonic contraversive ocular tilt reaction due to unilateral meso-diencephalic lesion. Neurology 40:1503–1509
11. Jones EG, Schward HD, Callahan PA (1986) Extent of ipsilateral representation in the ventral posterior medial nucleus of the monkey thalamus. Exp Brain Res 63: 310–320
12. Kinney FC (1978) An experimental study of the central gustatory pathways in the monkey, *Macaca mulatta* and *Cercopithecus aethiops*. J Hirnforsch 19:21–43
13. Motta G (1959) I centri corticali del gusto. Boll Sci Med 131:480–493
14. Nageotte J (1906) The pars intermedia or nervus intermedius of Wrisberg, and the bulbopontine gustatory nucleus in man. Rev Neurol Psychiatry 4:472–488
15. Nakajima Y, Utsumi H, Takahashi H (1983) Ipsilateral disturbance of taste due to pontine hemorrhage. J Neurol 229:133–136
16. Norgren R (1990) Gustatory system. In: Paxinos G (ed) The human nervous system. Academic Press, San Diego, pp 845–861
17. Schwartz HG, Roulac GE, Lam RL, O'Leary JL (1951) Organization of the fasciculus solitarius in man. J Comp Neurol 94:221–239
18. Travers SP, Pfaffmann C, Norgren R (1986) Convergence of lingual and palatal gustatory neural activity in the nucleus of the solitary tract. Brain Res 365:305–320

Subject Index

MIX
Papier aus verantwortungsvollen Quellen
Paper from responsible sources
FSC® C105338

If you have any concerns about our products,
you can contact us on
ProductSafety@springernature.com

In case Publisher is established outside the EU,
the EU authorized representative is:
Springer Nature Customer Service Center GmbH
Europaplatz 3, 69115 Heidelberg, Germany

Printed by Libri Plureos GmbH
in Hamburg, Germany